The Quantal Theory of Immunity

The Molecular Basis of
Autoimmunity and Leukemia

The Quantal Theory of Immunity

The Molecular Basis of
Autoimmunity and Leukemia

Kendall A. Smith

Cornell University, USA

World Scientific

NEW JERSEY • LONDON • SINGAPORE • BEIJING • SHANGHAI • HONG KONG • TAIPEI • CHENNAI

Published by

World Scientific Publishing Co. Pte. Ltd.

5 Toh Tuck Link, Singapore 596224

USA office: 27 Warren Street, Suite 401-402, Hackensack, NJ 07601

UK office: 57 Shelton Street, Covent Garden, London WC2H 9HE

Library of Congress Cataloging-in-Publication Data
Smith, Kendall A.
    The quantal theory of immunity : the molecular basis of autoimmunity and leukemia /
Kendall A. Smith.
      p. ; cm.
    Includes bibliographical references.
    ISBN-13: 978-981-4271-75-2 (hardcover : alk. paper)
    ISBN-10: 981-4271-75-6 (hardcover : alk. paper)
    1. Immune response. 2. Interleukin-2. 3. T cells. I. Title.
    [DNLM: 1. Autoimmunity--immunology. 2. Immune System--immunology. 3. Leukemia--
immunology. 4. Models, Theoretical. 5. Quantum Theory. QW 545 S653q 2010]
    QR186.S65 2010
    616.97'8--dc22

                                                                            2010025735

British Library Cataloguing-in-Publication Data
A catalogue record for this book is available from the British Library.

Typeset by Stallion Press
Email: enquiries@stallionpress.com

Printed in Singapore.

Dedication

To my heroes, Maurice Landy and Arthur Pardee,
all of the many trainees and colleagues,
and especially
to the next generation

Acknowledgements

Many thanks go to my dearest wife Lynn, who has supported me throughout the past 50 years, and to my colleagues who have been willing to read and criticize my efforts. Special thanks go to Doreen Cantrell and Ellis Reinherz for their critiques, and to The National Institute of Allergy and Infectious Diseases and to The Belfer Foundation for their continued support.

Prologue

When Burnet proposed the Clonal Selection Theory in 1957, he set the stage for immunologists for the next 50 years. However, at that time he could not begin to imagine the complexity of all of the cells that make up the immune system, and he could not dream of understanding any way in which molecules function to promote a systemic immune response. After the advent of cellular immunology of the 1960s and '70s, followed over the last three decades by a detailed identification and description of the myriad of molecules involved in regulating the behavior of the cells comprising the system, we now know *what* happens upon introduction of an antigen. However, we still have not discerned *how* the system or the cells are regulated, especially as to remain nonreactive to self-molecules yet to recognize and react with all molecules that are non-self. With the discovery around 1970 that there are two distinct types of lymphocytes, B cells and T cells, immunologists have focused on the notion that there are a myriad of separate lineages of cells that mature and function when called upon. Thus, we now speak of T helper (Th) cells and cytotoxic T (Tc) cells, as well as Th1 cells, Th2 cells and even Th17 cells. Moreover, recently the "Regulatory T cell" has come under investigation as a resurrection of the "Suppressor T cell" of 30 years ago. However, a unifying hypothesis of how immunity is regulated at the systemic, cellular and molecular levels has not been formulated. Accordingly, this volume is an attempt to build upon Burnet's prescient Clonal Selection Theory, and set the stage for the years to come, to propose how the critical molecules function to either promote or inhibit the cells from reacting to antigens, and to explain the systemic reactivity that is quantal, i.e. all-or-none, and represents what we recognize as immunity, or the exemption from disease.

Contents

Chapter 1

Introduction — The Evolution of Our Understanding of the Immune System

The immune system has fascinated scientists ever since Sir Edward Jenner demonstrated the quantal (all-or-none) nature of protection against smallpox in 1798, over 200 years ago. Although it had been known for centuries that some diseases never strike twice, smallpox among them, Jenner demonstrated for the first time that one could achieve such protection artificially and safely by exposure to an avirulent form of the disease, a practice that came to be known as vaccination (from the Latin *vaccinus*, from cows).[1] Working meticulously in his medical practice in the English countryside, Jenner tested the well-known observation that milkmaids were immune to smallpox; provided they had previously contracted a self-limited pox disease from the udders of cows. Thus, Jenner showed that immunity to smallpox could be transferred to individuals who had never been exposed to either cowpox or smallpox by inoculating them with pus from milkmaids who had cowpox.

Exactly how the vaccine changed the host to confer this immunity from such a serious disease (smallpox accounted for 10% of the mortality in the 18th century), remained obscure for the next 150 years, only becoming known following work carried out over the last 30–40 years to identify and understand the immune system. For almost a century, smallpox was thought to be a special case, and nobody attempted to prevent additional epidemic or endemic diseases via vaccination. During most of the 19th century contagious diseases were thought to be caused by an imbalance of the four bodily humors (i.e. blood, yellow bile, black bile, phlegm) first proposed by Hippocrates

(460–370 BC) two thousand years earlier. However, throughout the early and mid 19th century these ancient beliefs began to be questioned for the first time, and the germ theory of fermentation and putrefaction was formulated. Nevertheless, it was not until Louis Pasteur took up the debate in the late 1850s, repeating the earlier experiments of Schwann[2] and Cagniard-Latour,[3] that the "germ theory" became popularized and accepted as responsible for both fermentation and putrefaction, and also as a possible cause of contagious diseases.[4] Even so, the work of microbiology did not begin until another country doctor, Robert Koch, published his seminal experimental work in 1876, proving for the first time that a microbe actually causes the disease anthrax.[5,6]

Soon thereafter, Pasteur made the transition from chemistry to microbiology and then to immunology. In 1880, he revolutionized thought regarding immunity, by introducing the concept of attenuation of microbes using specialized cultivation methods, which in retrospect could never have attenuated the virulence of the bacteria he studied. Nevertheless, Pasteur extended Jenner's finding of vaccination to attenuated live bacterial vaccines, first for chicken cholera,[7] and then for anthrax of sheep and cattle.[8,9] Throughout Pasteur's studies, he maintained that only living organisms could confer protection via vaccination. Moreover, he was adamant that live organisms were required because "they depleted the host of vital trace nutrients" which were necessary for survival and multiplication of the organisms. Thus, the "father of immunology" was totally mistaken about how the immune system functions and how vaccination works.

In 1890, von Behring and Kitasato discovered antibody activity in the sera of immunized animals, thereby revealing that immunity occurs as an active process on the part of the host in response to exposure to foreign antigens.[10] Subsequently, during the first half of the 20th century, immunology focused on discerning the molecular nature of antibody activity, culminating in the 1950s with the demonstration by Tiselius and Kabat, as well as work by Porter and Edelman, that serum contains globular molecules, gamma globulins, which have the antibody activity, and that are comprised of two distinct chains (heavy and light chains).[11–13]

Despite the promise of vaccination to rid the world of such serious contagions as tuberculosis and all acute bacterial infections, more than 50 years elapsed before another vaccine was realized. It proved more difficult to attenuate bacteria than Pasteur had prophesized, although a live attenuated vaccine against Yellow Fever, which was subsequently found to be due to a virus, was accomplished in the 1930s.[14] After the influenza pandemic of 1918, when approximately 40–50 million people perished worldwide, basic research focused on flu, and ultimately a filterable agent (now known as a virus) was identified as the causative microbe in the 1930s, thereby excluding a bacterial origin.[15,16] Then, during the Second World War, clinical trials conducted by Jonas Salk, among others, showed that influenza could be prevented by inoculation of a formalin-fixed, killed vaccine preparation grown in chicken embryos (disproving Pasteur's dogma that only living vaccines could work). These methods have persisted unchanged.

In 1957 MacFarlane Burnet proposed the Clonal Selection Theory of Immunity, which has served as the foundation for immunology to the present time.[17] Burnet first proposed that lymphocytes are the primary cells responsible for immunity. At the time, plasma cells had been known to be associated with antibody formation for almost a decade, attributable to seminal work by Astrid Fagraeus,[18] but plasma cells were not yet known to be derived from lymphocyte precursors. Fagraeus mistakenly assigned them to be derived from reticulum cells (macrophages) of the spleen. In addition, Burnet proposed that lymphocytes recognize foreign antigenic molecules by virtue of cell surface receptors, a concept suggested originally by Paul Erlich 50 years earlier.

Burnet also proposed that once selected by antigen, lymphocyte clones undergo a proliferative expansion, which allows the secretion of large quantities of antibody molecules that facilitate removal of the offending antigen. Moreover, he suggested that the expanded clones of antigen-reactive cells could then mount a more rapid and greater immune response upon reintroduction of the same antigen, thereby providing a cellular basis for the phenomenon of "immunological memory," the basis for vaccination. Accordingly, the proliferation of lymphocytes stimulated by antigen became one of the axioms of immunity.

To explain the phenomenon of the inability of the immune system to react with self-molecules, Burnet further predicted that self-reactive cells are deleted during lymphocyte development. Accordingly, Burnet ascribed "self–non-self recognition" to be a function of the cells and their surface antigen receptors. However, the molecular mechanisms whereby these cellular processes occur were left unexplained and remained unapproachable.

In 1960, Peter Nowell provided the first demonstration that lymphocytes are capable of proliferating in response to mitogenic lectins,[19] and others soon extended this observation to specific antigens,[20,21] thereby founding cellular immunology. Subsequently, in the 1960s Jacques Miller[22] and Max Cooper and co-workers[23,24] showed that there are two distinct immune systems, one responsible for the generation of germinal centers, plasma cells and immunoglobulin (Ig) molecules, and another under the control of the thymus, responsible for delayed-type hypersensitivity (DTH), cell-mediated immunity (CMI), allograft rejection, and graft versus host disease (GvHD).

Two distinct types of lymphocytes were subsequently found to be responsible for humoral versus cellular immunity. B cells, detected by their expression of surface Ig (as predicted by Burnet) differentiate into antibody-forming plasma cells.[25] By comparison, T cells, which mature in the thymus, were found to be identifiable by their surface expression of theta (θ) antigen and their lack of surface expression of Ig.[26] Moreover, these surface markers allowed the removal of each subset via lysis with antibodies and complement, thereby permitting the dissection of their respective roles in the generation of immune responses.

Hozumi and Tonegawa then made the startling discovery that the genes encoding the antigen-binding variable region of the Ig molecules are distinct from those encoding the constant regions and that during lymphocyte development the two regions rearrange to join one another.[27] This finding was unprecedented and explained how DNA rearrangement could contribute to the tremendous diversity of antibody molecules. These findings were then confirmed and extended using DNA cloning and sequencing methods.[28–30]

Also by the 1960s, once the nature of antibody molecules became known, immunologists turned their attention to the molecular nature of antigens. Early studies by Benacerraf and co-workers revealed that there was a fundamental difference between the antigens reactive with antibody molecules, which could be shown to be small chemical entities, termed haptens, versus large proteins, such as bovine serum albumin (BSA), termed carriers, which were required to elicit a typical CMI response detected by a delayed-type hypersensitivity (DTH) cutaneous reaction.[31,32]

It could be shown that even artificial polypeptides, and thus neither self nor non-self peptides, could sensitize a host to prompt a DTH reaction,[33] and furthermore, a minimum of only six amino acids was required.[34] Thus, it could be assumed that proteins could contain many distinct epitopes. Even more perplexing, Benacerraf and colleagues found that both humoral and CMI are genetically regulated by immune response (Ir) genes.[35] McDevitt then mapped the Ir genes to the major histocompatibility complex (MHC) locus,[36,37] while Benacerraf and co-workers demonstrated that histocompatibility between T cells and B cells was found to be required for the generation of antibodies.[38] Then, Rosenthal and Shevach found that histocompatibility between macrophages and T cells was required to elicit an antigen-specific T cell proliferative response.[39] These findings led to the hypothesis that perhaps there existed two receptors in the immune system, one that recognized peptide antigens, and another that recognized self-receptors encoded by MHC genes, and that all of the cells that cooperate in the immune reaction had to have the same MHC-encoded molecules. However, when Zinkernagel and Doherty found that viral T cell cytolysis of virus-infected fibroblasts also required histocompatibility between "killer" T cells and virus-infected target cells, it appeared that the T cell antigen receptor (TCR) might simultaneously recognize both peptide epitopes and molecules encoded by MHC genes via a single receptor.[40]

Even so, the molecular nature of the MHC gene products that specified these genetic restrictions remained an enigma, as was the molecular nature of the T cell antigen receptor (TCR). Actually discovering the nature of these molecules became the Holy Grail of immunology. Using inbred (genetically identical) and congenic strains (genetically identical except for distinct chromosomal regions) of mice, Snell and others had mapped the chromosomal locus responsible for histocompatibility by performing skin grafts and tumor allotransplants in the 1940s.[41,42] Moreover, using antisera from immunized mice or from multiparous women, cell surface molecules were identified as histocompatibility leukocyte antigens (HLA) by Dausset, Strominger, Hood and others.[43-46] However, it was not until molecular genetics could be applied by Hood, Steinmetz and others in the early 1980s that the tremendous polymorphism of the MHC genes was appreciated (for review see Ref. 47).

The diversity of antigen recognition had led Burnet to predict that the "Clonal Selection Theory could never be tested experimentally unless in vitro culture methods could be developed that allowed for the creation of pure clones of lymphocytes."[48] Kohler and Milstein first satisfied such a condition for antibody-forming cells in 1975, by creating monoclonal antibody secreting somatic cell hybrids (hybridomas) between malignant proliferating plasma cells and antibody producing B cells from splenocytes.[49]

With regard to T cells, in 1965 two groups reported that medium conditioned by alloantigen-stimulated lymphocytes contained mitogenic factors.[50,51] Then, for the next decade numerous reports appeared of mitogenic activities in the conditioned media from leukocyte cultures, some thought to be derived from macrophages and others from lymphocytes.[52-58] Also, in the mid-1970s several groups showed that repetitive alloantigen stimulation could promote the growth of T cells in culture for several months.[59-61] Subsequently, Morgan and co-workers showed that lymphocyte-conditioned media could support the long-term culture of T cells from bone marrow, which suggested that the cultured cells might be derived from immature T cell precursors.[62]

Because the prevailing immunological dogma indicated that only antigen was capable of promoting T cell proliferation, it seemed improbable that one could use the lymphocyte-conditioned medium to support long-term antigen-specific T cell proliferation. However, the very first experiments were successful,[63] as were cloning experiments that established the first monoclonal antigen-specific cytolytic T cells in 1979.[64] Accordingly, monoclonal hybridomas cells secreting antigen-specific monoclonal antibodies, and monoclonal T cells cytolytic for specific antigens, provided the data that proved Burnet's Clonal Selection Theory for both B cells and T cells. In addition, the capacity to grow and study antibody producing hybridoma clones, and antigen-specific functional T cell clones, ushered in the era of molecular immunology, which began in 1980 and is still ongoing.

References

1. Jenner, E. (1798) *An Inquiry into the Causes and Effects of Variolae Vaccinae, a Disease Discovered in Some Western Counties of England.* Sampson Low. London
2. Schwann, T. (1837) Preliminary report on experiments concerning alcoholic fermentation and putrefaction. *Annalen der Physik und Chemie* **41**:184–193.
3. Cagniard-Latour, C. (1838) Memoire on alcohol fermentation. *Annales de Chimie et de Physique* **68**:206–222.
4. Pasteur, L. (1857) Mémoire sur la fermentation appelée lactique. (Extrait par l'auteur). *Comptes Rendus des Seances de L'Académie des Sciences* **45**:913–916.
5. Koch, R. (1876) Die aetiologie der milzbrand-krankheit, begrundet auf die entwicklungsgeschichte des bacillus antracis. *Beitrage zur Biologie der Pflanzen* **2**:277–310.
6. Pasteur, L., Joubert, and Chamberland. (1878) La théorie des germes et ses applications à la medicine et à la chirurgie. *Comptes Rendus Hebdomadaires des Séances de l'Académie des Sciences* **86**:1037–1043.
7. Pasteur, L. (1880) Sur les maladies virulentes, et en particulier sur la maladie appelée vulgairement cholera des poules. *Comptes Rendus Hebdomadaires des Séances de l'Académie des Sciences* **90**:248–249.
8. Pasteur, L. (1881) Compte rendu sommaire des experiences faites à Pouilly-Le-Fort, près de Melun, sur la vaccination charnonneuse (avec la collaboration de MM. Chamberland et Roux). *Compte Rendus Acad. Sci.* **XCII**:1378–1383.
9. Pasteur, L., Chamberland, and Roux. (1881) De l'attenuation des virus et de leur retore à la virulence. *Comptes Rendus des Séances de L'Académie des Sciences* **92**:430–435.

10. Behring, E., and Kitasato, S. (1890) Concerning development of diphtheria immunity and tetanus immunity in animals. *German Medical Weekly*.

11. Tiselius, A., and Kabat, E. (1939) An electrophoretic study of immune sera and purified antibody preparations. *J. Exp. Med.* **65**:119–131.

12. Porter, R.R. (1959) The hydrolysis of rabbit gamma globulin and antibodies with crystalline papain. *Biochem. J.* **73**:119–138.

13. Edelman, G.M. (1959) Dissociation of gamma globulin. *J. Am. Chem. Soc.* **81**:3155–3170.

14. Theiler, M., and Smith, H. (1936) The use of Yellow Fever Virus modified by *in vitro* cultivation for human immunization. *J. Exp. Med.* **65**:787–800.

15. Shope, R. (1931) Swine influenza: I Experimental transmission and pathology. *J. Exp. Med.* **54**:349–359.

16. Smith, W., Andrews, C., and Laidlaw, P. (1933) A virus obtained from influenza patients. *Lancet* **2**:66–68.

17. Burnet, F.M. (1957) A modification of Jerne's theory of antibody production using the concept of clonal selection. *Aust. J. Sci.* **20**:67–77.

18. Fagraeus, A. (1948) The plasma cellular reaction and its relation to the formation of antibodies *in vitro*. *J. Immunol.* **58**:1–13.

19. Nowell, P.C. (1960) Phytohemagglutinin: an initiator of mitosis in cultures of normal human leukocytes. *Cancer Research* **20**:462–468.

20. Hirschhorn, K., Bach, F., Kolodny, R., Firschein, I., and Hashem, N. (1963) Immune response and mitosis of human peripheral blood lymphocytes *in vitro*. *Science* **142**:1185–1187.

21. Bain, B., and Lowenstein, L. (1964) Genetic studies on the mixed leukocyte reaction. *Science* **145**:1315–1316.

22. Miller, J. (1962) Effect of neonatal thymectomy on the immunological responsiveness of the mouse. *Proc. Roy. Soc. London Series B* **156**:415–428.

23. Cooper, M., Peterson, R., and Good, R. (1965) Delineation of the thymic and bursal lymphoid systems in the chicken. *Nature* **205**:143–146.

24. Cooper, M., Peterson, R., South, M., and Good, R. (1966) The functions of the thymus system and the bursa system in the chicken. *J. Exp. Med.* **123**:75–102.

25. Raff, M., Sternberg, M., and Taylor, R.B. (1970) Immunoglobulin determinants on the surface of mouse lymphoid cells. *Nature* **225**:553–555.

26. Raff, M. (1969) Theta isoantigen as a marker of thymus-derived lymphocytes in mice. *Nature* **224**:378–379.

27. Hozumi, N., and Tonegawa, S. (1976) Evidence for somatic rearrangement of immunoglobulin genes coding for variable and constant regions. *Proc. Natl. Acad. Sci. USA* **73**:3628–3632.

28. Tonegawa, S., Brack, C., Hozumi, N., and Schuller, R. (1977) Cloning of an immunoglobulin variable region gene from mouse embryo. *Proc. Natl. Acad. Sci. USA* **74**:3518–3523.

29. Bernard, O., Hozumi, N., and Tonegawa, S. (1978) Sequences of mouse immunoglobulin light chain genes before and after somatic changes. *Cell* **15**:1133–1139.
30. Seidman, J., Edgell, M., and Leder, P. (1978) Immunoglobulin light chain structural gene sequences cloned in a bacterial plasmid. *Nature* **271**:582–586.
31. Benacerraf, B., and Gell, P. (1959) Studies on hypersensitivity. I. Delayed and Arthus-type skin reactivity to protein conjugates in guinea pigs. *Immunol.* **2**:53–63.
32. Gell, P., and Benacerraf, B. (1959) Studies on hypersensitivity. II. Delayed hypersensitivity to denatured proteins in guinea pigs. *Immunol.* **2**:64–70.
33. Kantor, F.S., Ojeda, A., and Benacerraf, B. (1963) Studies on artificial antigens I. Antigenicity of DNP-polylysine and DNP copolymer of lysine and glutamic acid in guinea pigs. *J. Exp. Med.* **117**:55–64.
34. Schlossman, S., Ben-Efraim, S., Yaron, A., and Sober, H. (1966) Immunochemical studies on the antigenic determinants required to elicit delayed and immediate hypersensitivity reactions. *J. Exp. Med.* **123**: p1083–p1095.
35. Green, I., Paul, W.E., and Benacerraf, B. (1966) The behavior of hapten-poly-L-lysine conjugates as complete antigens in genetic responder and as haptens in non-responder guinea pigs. *J. Exp. Med.* **123**:859–879.
36. McDevitt, H.O., and Tyan, M.L. (1968) Genetic control of the antibody response in inbred mice: transfer of response by spleen cells and linkage to the major histocompatibility (H2) locus. *J. Exp. Med.* **128**:1–11.
37. McDevitt, H.O., and Chinitz, A. (1969) Genetic control of the antibody response: relationship between immune response and histocompatibility (H-2) type. *Science* **163**:273–279.
38. Katz, D., Hamaoka, T., and Benacerraf, B. (1973) Cell interactions between histoincompatible T and B lymphocytes II. Failure of physiologic cooperative interactions between T and B lymphocytes from allogeneic donor strains in humoral response to hapten-protein conjugates. *J. Exp. Med.* **137**:1405–1418.
39. Rosenthal, A., and Shevach, E. (1973) Function of macrophages in antigen recognition by guinea pig T lymphocytes. I. Requirement for histocompatible macrophages and lymphocytes. *J. Exp. Med.* **138**:1194–1212.
40. Zinkernagel, R., and Doherty, P. (1974) Restriction of *in vitro* T cell-mediated cytotoxicity in lymphocytic choriomeningitis within a syngeneic or semiallogeneic system. *Nature* **248**:701–702.
41. Snell, G. (1948) Methods for the study of histocompatibility. *J. Genetics* **49**:87–108.
42. Gorer, P., Lyman, S., and Snell, G. (1948) Studies on the genetic and antigenic basis of tumor transplantation. Linage between a histocompatibility gene and 'Fused' in mice. *Proc. Royal Soc. London Series B* **135**:499–505.
43. Dausset, J. (1958) Iso-leuko-antibodies. *Acta Haematol.* **20**:156–166.

44. Sanderson, A., and Batchelor, J. (1968) Transplantation antigens from human spleens. *Nature* **219**:184–187.

45. Cresswell, P., Turner, M., and Strominger, J. (1973) Papain solubilized HL-A antigens from cultured human lymphocytes contain two peptide fragments. *Proc. Natl. Acad. Sci. USA* **70**:1603–1607.

46. Silver, J., and Hood, L. (1976) Structure and evolution of transplantation antigens: partial amino acid sequences of H-2K and H2-D alloantigens. *Proc. Natl. Acad. Sci. USA* **73**:599–603.

47. Steinmetz, M., and Hood, L. (1983) Genes of the Major Histocompatibility Complex in mouse and man. *Science* **222**:727–733.

48. Burnet, F.M. (1959) *The Clonal Selection Theory of Acquired Immunity.* Cambridge University Press. Cambridge. 49–80 pp.

49. Kohler, G., and Milstein, C. (1975) Continuous culture of fused cells secreting antibody of predefined specificity. *Nature* **256**:495–499.

50. Kasakura, S., and Lowenstein, L. (1965) A factor stimulating DNA synthesis derived from the medium of leukocyte cultures. *Nature* **208**:794–795.

51. Gordon, J., and MacLean, L.D. (1965) A lymphocyte-stimulating factor produced *in vitro. Nature* **208**:795–796.

52. Bach, F., Alter, B., Solliday, S., Zoschke, D., and Janis, M. (1970) Lymphocyte reactivity *in vitro* II. Soluble reconstituting factor permitting response of purified lymphocytes. *Cell. Immunol.* **1**:219–227.

53. Hoffman, M., and Dutton, R. (1971) Immune response restoration with macrophage culture supernatants. *Science* **172**:1047–1048.

54. Gery, I., Gershon, R.K., and Waksman, B. (1972) Potentiation of the T-lymphocyte response to mitogens. *J. Exp. Med.* **136**:128–142.

55. Gery, I., and Waksman, B.H. (1972) Potentiation of the T-lymphocyte response to mitogens: the cellular source of potentiating mediators. *J. Exp. Med.* **136**:143–155.

56. Schimpl, A., and Wecker, E. (1972) Replacement of T cell function by a T cell product. *Nature New Biol.* **237**:15–17.

57. Plate, J. (1976) Soluble factors substitute for T-T-cell collaboration in the generation of T-killer lymphocytes. *Nature* **260**:329–331.

58. Delovitch, T., and McDevitt, H. (1977) *In vitro* analysis of allogeneic lymphocyte interaction. I. Characterization of an Ia-positive helper factor-allogeneic effect factor. *J. Exp. Med.* **146**:1019–1026.

59. Ben-Sasson, S., Paul, W., Shevach, E., and Green, I. (1975) *In vitro* selection and extended culture of antigen-specific T lymphocytes. I. Description of selection procedure and initial characterization of selected cells. *J. Exp. Med.* **142**:90–105.

60. Svedmyr, E. (1975) Long-term maintenance *in vitro* of human T cells by repeated exposure to the same stimulator cells. *Scand. J. Immunol.* **4**:421–427.

61. Dennert, G., and De, R.M. (1976) Continuously proliferating T killer cells specific for H-2b targets: selection and characterization. *J. Immunol.* **116**: 1601–1606.
62. Morgan, D.A., Ruscetti, F.W., and Gallo, R. (1976) Selective *in vitro* growth of T lymphocytes from normal human bone marrows. *Science* **193**:1007–1008.
63. Gillis, S., and Smith, K.A. (1977) Long term culture of tumour-specific cytotoxic T cells. *Nature* **268**:154–156.
64. Baker, P.E., Gillis, S., and Smith, K.A. (1979) Monoclonal cytolytic T-cell lines. *J. Exp. Med.* **149**:273–278.

Chapter 2

Molecular Immunology

Using monoclonal T cells, it was possible to create a rapid unambiguous *in vitro* bioassay for the mitogenic activity in the lymphocyte-conditioned media, termed T cell growth factor (TCGF), and to then identify its molecular characteristics.[1] Lymphocyte-conditioned media was found to contain a single 15 kDa glycoprotein that was responsible for long-term T cell growth.[2] Moreover, using the bioassay, enough TCGF protein could be purified to generate MoAbs that could be used as immunoaffinity reagents to purify large amounts (i.e. milligrams) of homogeneous protein in a single step.[3] In addition, biosynthetically radiolabeled TCGF was purified and used to demonstrate the first cytokine receptor, which had all of the characteristics of a classic hormone receptor including high affinity, ligand specificity, target cell specificity and saturation binding.[4] Moreover, the TCGF radioreceptor assay was used to identify the first MoAb reactive with a cytokine receptor.[5]

The monoclonal cytolytic T cells also could be used to test additional mitogenic factors. Thus, it was found that lymphocyte activating factor (LAF), which had been identified as a macrophage-derived activity mitogenic for thymocytes,[6,7] in fact could not support the proliferation of cloned T cells. Experiments showed that LAF mitogenic activity was due to its capacity to augment TCGF production by T cells.[8,9] Accordingly, these data provided the scientific rationale for the "interleukin" nomenclature, a term meant to designate molecules that carry messages "between leukocytes." Because LAF worked upstream of TCGF, it was named IL-1, while TCGF became IL-2, thereby anticipating additional interleukins yet to be discovered.

The use of monoclonal antigen-specific T cells by Schwartz and Paul and their co-workers proved unequivocally that individual T cells recognize small synthetic peptide epitopes associated with Ia antigens specified by Ir genes of the MHC Class II region, and that both the restricting Ir gene products and the peptide epitope had to be presented by the same APC.[10] These experiments were the first to prove that the Ir gene products are actually the Ia antigens. Moreover, because Ia-reactive MoAbs blocked the proliferative response of cloned T cells, it could be concluded that "suppressor T cells" were not responsible for the anti-Ia suppression, a popular theory at the time. Instead, the Ia-reactive MoAbs bound to the molecules on the surface of the APC. Even so, the structure of the Ia antigens and how the T cell clone recognized both peptide epitopes and Ia antigens remained obscure. Accordingly, the "two T cell receptors versus one T cell receptor" hypotheses needed to be resolved to explain how T cells recognized antigens "in the context" of HLA molecules.

Reinherz and co-workers used lymphocyte-conditioned media containing IL-2 activity to generate the first antigen-specific human cytolytic T cell clones, and then used these clones to raise T cell clone-specific murine MoAb.[11] These "clonotypic" MoAbs were then shown to block antigen-specific proliferation *in vitro*, and to precipitate a 90 kDa disulfide-linked heterodimer of two molecules of ~45 kDa from the cell surface. This was the first demonstration of T cell antigen receptor (TCR) proteins. They also showed that the signaling components of the TCR consisted of a complex of smaller invariant proteins, designated CD3,[12] which are present on all T cells, and that two other T cell surface proteins, CD4 and CD8, provide the components of the TCR complex that ensured MHC Class restriction of antigen recognition.[13,14] Accordingly, CD4$^+$ T cells interact with specific antigen together with MHC Class II, while CD8$^+$ T cells interact with antigen only together with MHC Class I.

With the availability of cloned T cells and clonotypic MoAbs that could be used to identify the TCR/CD3 complex unambiguously, it was possible to show that triggering the TCR promotes cell cycle "competence," defined in cell cycle terms as G_0-G_1 transition, and in molecular terms as the expression of the genes encoding IL-2 and

IL-2 receptors (IL-2R).[15] However, cell cycle "progression," defined as G_1-S-phase transition, was found to be solely dependent upon the IL-2/IL-2R interaction.[16]

Subsequently, Stephen Hedrick and Mark Davis and co-workers,[17] as well as Tak Mak and co-workers,[18] isolated the first of the four TCR chain cDNAs, which led to the discovery that the TCR belongs to the immunoglobulin (Ig) super family, and is organized in the genome in a fashion similar to the genes encoding antibody molecules. Just like the Ig genes, the TCR genes could be found to generate diversity by undergoing rearrangement and recombination. Subsequent work by Davis and others showed that there are four chains that can form TCRs, the α and β chains form the most common class of TCRs and the γ and δ chains form a second, less common class of TCRs (reviewed in Ref. 19).

Strominger's group culminated their painstaking biochemical purification of HLA molecules from 200 L of cultured lymphoblastoid cell lines by generating 3–4 mg of Class I MHC-encoded A2 molecules. Bjorkman worked with Strominger's and Wiley's groups to provide three-dimensional crystal structures of these molecules that could be analyzed through X-ray diffraction to yield the much needed view of how the peptide epitope fits into a groove in the MHC-encoded molecules.[20,21] Thus, these data indicated that one TCR comprised of two covalently-linked chains recognizes both the peptide epitope and the HLA molecules, essentially proving the "one TCR hypothesis." However, they still did not yield the information necessary to understand how this event triggers the cell to produce IL-2 and to express IL-2Rs. Moreover, exactly how T cells are stimulated to proliferate via the IL-2/IL-2R interaction, and to differentiate into "effector cells" capable of secreting cytokines and cytolytic molecules was not known.

A great deal of information was accumulated regarding the early intracellular biochemical events triggered by the TCR. Intracellular kinase cascades triggered by the TCR converge on three families of transcription factors (TF), which include the activating protein-1 (AP-1), nuclear factor of activated T cells (NF-AT), and the Rel (NF-κB) families. Early experiments revealed how these three TFs combine to form an "enhancesome" at the IL2 promoter, and how all three members must be present to form a stable macromolecular complex capable of activating transcription.[22,23] Even so, the dynamics of these molecular activities have been difficult to discern, primarily because analyses have been performed at the cell population level, and it is not easy to relate the number of TCRs triggered over time and the TF activation of IL2 gene expression. Moreover, so many genes are activated during the first several hours after the reaction is initiated, cause versus effect relationships have not been forthcoming. Therefore, we now know "what happens" when a peptide antigen is recognized by naive T cells, but we do not know how and why things happen as they do.

References

1. Gillis, S., Ferm, M.M., Ou, W., and Smith, K.A. (1978) T cell growth factor: parameters of production and a quantitative microassay for activity. *J. Immunol.* **120**:2027–2032.
2. Robb, R.J., and Smith, K.A. (1981) Heterogeneity of human T-cell growth factor(s) due to variable glycosylation. *Mol. Immunol.* **18**:1087–1094.
3. Smith, K.A., Favata, M.F., and Oroszlan, S. (1983) Production and characterization of monoclonal antibodies to human interleukin 2: strategy and tactics. *J. Immunol.* **131**:1808–1815.
4. Robb, R.J., Munck, A., and Smith, K.A. (1981) T cell growth factor receptors: quantitation, specificity, and biological relevance. *J. Exp. Med.* **154**:1455–1474.
5. Leonard, W.J., Depper, J.M., Uchiyama, T., Smith, K.A., Waldmann, T.A., and Greene, W.C. (1982) A monoclonal antibody that appears to recognize the receptor for human T-cell growth factor; partial characterization of the receptor. *Nature* **300**:267–269.
6. Gery, I., Gershon, R.K., and Waksman, B. (1972) Potentiation of the T-lymphocyte response to mitogens. *J. Exp. Med.* **136**:128–142.
7. Gery, I., and Waksman, B.H. (1972) Potentiation of the T-lymphocyte response to mitogens: the cellular source of potentiating mediators. *J. Exp. Med.* **136**:143–155.
8. Smith, K.A., Gilbride, K.J., and Favata, M.F. (1980) Lymphocyte activating factor promotes T-cell growth factor production by cloned murine lymphoma cells. *Nature* **287**:853–855.
9. Smith, K.A., Lachman, L.B., Oppenheim, J.J., and Favata, M.F. (1980) The functional relationship of the interleukins. *J. Exp. Med.* **151**:1551–1556.

10. Sredni, B., Matis, L., Lerner, E., Paul, W., and Schwartz, R. (1981) Antigen-specific T cell clones restricted to unique F1 major histocompatability determinants; inhibition of proliferation with a monoclonal anti-Ia antibody. *J. Exp. Med.* **153**:677–693.

11. Meuer, S.C., Fitzgerald, K.A., Hussey, R.E., Hodgdon, J.C., Schlossman, S., and Reinherz, E.L. (1983) Clonotypic structures involved in antigen-specific human T cell function. Relationship to the T3 molecular complex. *J. Exp. Med.* **157**:705–719.

12. Reinherz, E.L., Hussey, R.E., and Schlossman, S.F. (1980) A monoclonal antibody blocking human T cell function. *Eur. J. Immunol.* **10**:758–762.

13. Reinherz, E.L., Kung, P.C., Goldstein, G., and Schlossman, S.F. (1979) Separation of functional subsets of human T cells by a monoclonal antibody. *Proc. Natl. Acad. Sci. USA* **76**:4061–4065.

14. Reinherz, E.L., and Schlossman, S.F. (1980) The differentiation and function of T lymphocytes. *Cell* **4**:821–827.

15. Meuer, S.C., Hussey, R.E., Cantrell, D.A., Hodgen, J.C., Schlossman, S.F., Smith, K.A., and Reinherz, E.L. (1984) Triggering the T3-Ti antigen-receptor complex results in clonal T cell proliferation through an interleukin 2-dependent autocrine pathway. *PNAS* **81**:1509–1513.

16. Cantrell, D.A., and Smith, K.A. (1983) Transient expression of interleukin 2 receptors. Consequences for T cell growth. *J. Exp. Med.* **158**:1895–1911.

17. Hedrick, S., Cohen, D., Nielson, E., and Davis, M.M. (1984) Isolation of cDNA clones encoding T cell-specific membrane associated proteins. *Nature* **308**:149–155.

18. Yanagi, Y., Yoshikai, Y., Leggett, K., Clark, S., Aleksander, I., and Mak, T. (1984) A human T cell-specific cDNA clone encodes a protein having extensive homology to immunoglobulin chains. *Nature* **308**:145–149.

19. Davis, M., and Bjorkman, P. (1988) T cell antigen receptor genes and T cell recognition. *Nature* **334**:395–402.

20. Bjorkman, P.J., Saper, M.A., Samaoui, B., Bennett, W.S., Strominger, J.L., and Wiley, D.C. (1987) Structure of the human class I histocompatability antigen, HLA-A2. *Nature* **329**:506–511.

21. Bjorkman, P., Saper, M., Samaraoui, B., Bennet, W.S., Strominger, J., and Wiley, D. (1987) The foreign antigen binding site and T cell recognition regions of class I histocompatability antigens. *Nature* **329**:512–518.

22. Garrity, P.A., Chen, D., Rothenberg, E.V., and Wold, B.J. (1994) Interleukin 2 transcription is regulated *in vivo* at the level of coordinated binding of both constitutive and regulated factors. *Mol. Cell. Bio.* **14**:2159–2169.

23. Rothenberg, E.V., and Ward, S.B. (1996) A dynamic assembly of diverse transcription factors integrates activation and cell-type information for interleukin 2 gene regulation. *Proc. Natl. Acad. Sci. USA* **93**:9358–9365.

Chapter 3

The Problem — Understanding How Molecules Direct the Behavior of Cells Comprising the Immune System

Burnet correctly based the behavior of the immune system on decisions made by individual cells comprising the system, and also correctly predicted that each cell reacts with an environmental antigen via a clone-specific cell surface receptor. However, Burnet had no way of knowing how the antigen receptor molecules actually "recognize" a foreign antigen molecule, and what molecules are involved in transferring the information imparted by antigen-receptor interaction at the cell surface to the cellular interior. Moreover, how this information becomes translated into determining the tempo, magnitude and duration of a systemic immune response simply could not be discerned in 1959 because the molecules involved had not yet been discovered.

Consequently, for the past 50 years, immunologists have *assumed* that an immune response rests entirely with the presence or absence of an antigen. Thus, the thinking has been that the cells comprising the immune system are quiescent until the antigen is introduced, after which the cells become activated to proliferate and differentiate, whereupon they clear the antigen from the system. Subsequently, there is no longer any positive driving antigenic force, and the cells consequently return once again to a resting, quiescent state. However, the primacy of antigen as the ultimate regulator of immune responsiveness came into question once it was realized that "leukocytotrophic hormones," a.k.a. interleukins or cytokines, are actually the molecules responsible for carrying out the antigen-initiated cellular changes.

Axiomatic in this regard is that the tempo, magnitude and duration of proliferative clonal expansion and subsequent differentiation that occurs during an immune response is directed not by the antigen *per se*, but by the leukocytotrophic hormones that are produced after antigenic stimulation. Since IL-2 mediates T cell cycle progression, the molecular determinants of IL-2 gene expression and IL-2R gene expression become paramount for the generation of a T cell immune response. With regard to B cells, additional interleukins such as IL-4 and IL-6, as well as ligands and receptors of the tumor necrosis factor (TNF) family, serve a similar role. Thus, although the immune system is initially awakened by the antigen-receptor interaction with foreign environmental molecules that gain entry to the internal milieu, the immune system has evolved endogenous leukocytotrophic *hormones* that subsequently determine how the system reacts to the recognition of foreign molecules.

Especially within the past 25 years, experiments by many individuals have elucidated what molecules are involved in antigen "recognition" by TCRs and BCRs. However, how the molecules interact to lead to a productive response on the part of the cell is only just now beginning to be unraveled. Accordingly, ligand-receptor *signaling* at the cell surface becomes translated into a quantal response at the level of enhancesomes controlling specific individual genes. It is clear that more than one antigen-receptor interaction is required to mediate a recognition event, but it has not been discerned exactly how many antigen-receptor molecular interactions must occur, or over what time interval, to lead to meaningful gene expression. In other words, not only does one need to know the *biochemistry* of the antigen-receptor recognition and signaling events, but one must also understand the *biophysics* of the molecular interactions. Moreover, the molecular determinants of gene expression must be connected to the signals emanating from the surface antigen-receptor interactions.

Once antigen "recognition" occurs, the internal leukocytotrophic hormone systems take control of the immune response, so that it is *internally* regulated, as are all of the other recognized body systems.

Exactly the same kinds of biochemical and biophysical parameters that determine antigen recognition and signaling must now be discerned for cytokine-receptor recognition and signaling. Thus far, much has been uncovered as to the biochemical parameters employed by cytokine families and their respective receptors. However, we are only beginning to ascertain how complicated the individual cellular responses are to a single cytokine-receptor interaction. In addition, we are only beginning to uncover the many feedback regulatory controls that govern cytokine gene expression and signaling.

Accordingly, this volume is focused on building upon and extending Burnet's Clonal Selection Law of Immunity, to reduce the problems from the systemic to the cellular, and finally to the molecular levels, so that new approaches to abnormalities of the molecules of the immune system that lead to disease will be possible.

Chapter 4

The Quantal Theory of Immunity

Any new theory that attempts to explain how the immune system functions to discriminate between non-self foreign antigens versus self auto-antigens must first obey Burnet's Law of Clonal Selection and expansion at the level of each individual cell. In addition, any new theory must provide a molecular explanation for the behavior of the individual cells comprising the system. Most importantly, the theory must account for the reactivity of the whole immune system to an antigen, whether non-self or self. This should be possible, now that we know the molecular nature of the receptors for antigens as well as the molecular nature of antigens. In addition, we have identified the leukocytotrophic hormone molecules as well as their receptor molecules which take over the control of immune reactivity after the introduction of an antigen.

In biology, the term *quantal* refers to a response that either occurs or does not, i.e. all-or-none, or as something existing in only one of two possible states, e.g. dead or alive. As it applies to an immune response, we consider it natural that immunity is quantal. We speak of individuals as being immune or not, yet the science of immunology is focused on trying to explain how this quantal systemic state comes about. Brent and Medawar first proposed a "quantal theory of the immune response" to account for strong and weak delayed-type hypersensitivity reactions to injections of allogeneic cells.[1] Thus, they attributed the magnitude of the DTH response to be due "solely to the number of cells engaged in an immunological performance. According to this view, any one lymphocyte either does or does not react (hence "quantal"): there is no such thing as a sensitized cell in this situation — only a sensitized population." What Brent and Medawar did not consider was that the difference between a sensitized and unsensitized population is the number or proportion of antigen-reactive cells.

At this time in the evolution of immunological theory, it is axiomatic that we must now examine and explain how each individual cell makes the decision to react or not, and how the critical molecules involved actually function to bring about the quantal cellular response.

In other branches of biology, examples of quantal responses are familiar, e.g. the contraction of a muscle as a result of stimulation at the neuromuscular junction, or cell fate determination so well known in embryogenesis. These phenomena are really a variation of Erwin Schrödinger's famous "cat experiment" in which he proposed to demonstrate the stochastic (probabilistic) nature of quantum mechanics.[2] Another way of demonstrating the phenomenon is to expose a population of insects to an insecticide. Thus, quantal phenomena like life or death at the macroscopic level of the whole individual would be seen as stochastic when viewed at the population or group level. That is, some individuals will succumb before others, with most distributed about the mean for the population, and the differences between individuals of the population will appear probabilistic. One of the difficulties with immunology is that many of our experiments have been performed with populations of animals or cells, yet what we really need and want to know is what determines the behavior of individuals, both animals and cells, and to discern whether their behavior is probabilistic or deterministic.

The Quantal Theory of Immunity states that individual cells of the immune system recognize and react to antigens (both non-self and self) by proliferating and differentiating into effector cells in an all-or-none (quantal) fashion.[3,4] Furthermore, the Quantal Theory states that individual cells make this quantal decision only after "counting" the number of triggered antigen receptors, which ultimately determines the quantal expression of the IL-2 and IL-2R genes. After antigen receptors have been triggered, the quantal decision of the cell to progress through the G_1 phase of the cell cycle to the S-phase and undergo cytokinesis, which is the basis of clonal expansion, is determined by a crucial number of IL-2/IL-2R interactions, which the cell counts. The premise of the Quantal Theory is that these same principles underlie all of the cell fate decisions in the immune system, i.e. cellular differentiation to effector functions, conversion to long-lived memory cells, as well as the decisions to become anergic and thus suppressive, or to undergo apoptosis.

References

1. Brent, L., and Medawar, P. (1967) Cellular immunity and the homograft reaction. *Brit. Med. Bull.* **23**:55–60.
2. Schrodinger, E. (1935) Cat experiment. *http://www.mtnmath.com/faq/meas-qm-3.html.*
3. Smith, K. (2004) The quantal theory of how the immune system discriminates between "self and non-self." *Med. Immunol.* **3**:3.
4. Smith, K. (2006) The quantal theory of immunity. *Cell. Res.* **16**:11–19.

Chapter 5

The Variability of Cell Cycle Progression and the Competence and Progression Phases of the Cell Cycle

Before focusing on the critical molecules responsible for lymphocyte activation and proliferation, it is helpful to understand the evolution of our knowledge of the regulation of the proliferation of all somatic cells. Beginning in 1932 with studies on bacterial cell growth, numerous experiments with many different cell types over the following 50 years revealed that the cell cycle times of individual cells within a population follow a normal distribution when examined as a function of the division rate (termed the rate-normal distribution) (reviewed in Ref. 1). Thus, some cells have very rapid generation times, while others are slower, with the majority of cells within the population distributed about the mean. This distribution is also a log-normal distribution, since rate is a reciprocal of the time required to progress through the cell cycle. Data from experiments with the flagellated protozoan *Euglena gracilis*, often quoted in the cell cycle literature, are illustrated in Fig. 5.1 to emphasize the variability in the rate of growth of individual cells within a population.[2]

Studies conducted with either synchronized or asynchronously proliferating cell populations indicated that the variability in the cell cycle transit times among the individual cells comprising the population occurs in the pre-replicative phases of the cell cycle (i.e. G_0-G_1), as the replicative phases (i.e. S, G_2, M) remain relatively constant for a given cell population.[3] For example, studies using time-lapse cinemicrography to monitor the G_1 time from the completion of anaphase

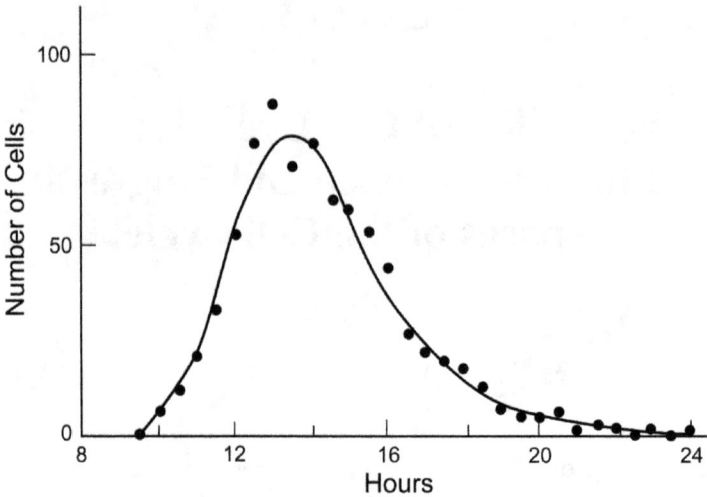

Figure 5.1:   Cell cycle lengths. Time-lapse cinematography was used to determine interdivision times of exponentially growing *Euglena gracilis*. The number of cells with division times within each half-hour interval are shown on the y-axis versus intermitotic time on the x-axis. (Redrawn from: Cook, J. and Cook, B. 1962. *Exp. Cell Res.* **28**:524–534.)

to the onset of DNA synthesis of human amniotic cells is shown in Fig. 5.2. It is evident that the G_1 times of individual cells are highly variable. Like runners in the New York Marathon, some cells progress through G_1 rapidly, entering S-phase (the finish line) within only six hours while others required more than 20 hours to complete G_1. When these data are plotted using probit analysis[4] and the rates of each cell traversing G_1 are plotted as shown in Fig. 5.3, in the case of the total cell cycle the G_1 rates are normally distributed. This can be seen graphically from Fig. 5.3.

Also, the duration of the cell cycle is not genetically determined nor is it passed on at division, since the correlations of cell cycle times for mother–daughter cells are generally poor, even for cloned cell populations. However, cycle times of sister cells do show a positive correlation, an indication that sibling cells inherit some property that contributes to similar cell cycle transit times for the next cycle, which

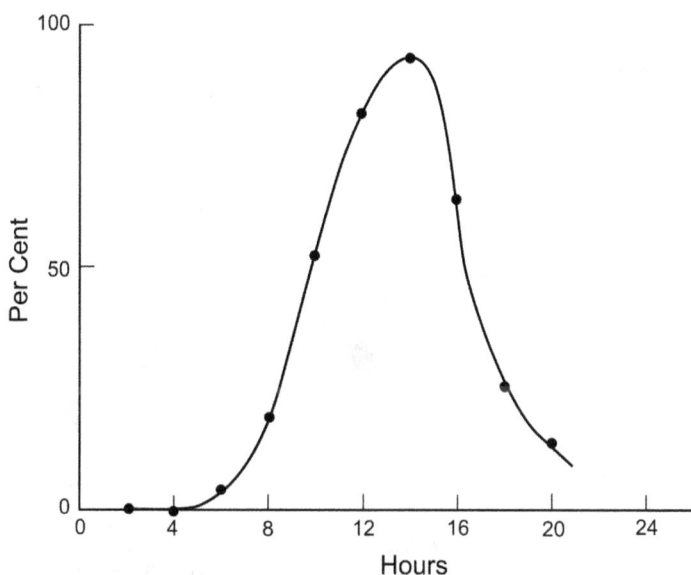

Figure 5.2: G₁ intervals of human amniotic cells. Distributions of the percentages of cells that had begun to incorporate tritiated thymidine into newly synthesized DNA after short labeling intervals. (Redrawn from: Sisken, J. and Morasca, L. 1965. *J. Cell Biol.* **25**:179–189.)

then disappears by the subsequent cell cycle. Consequently, there is no selection for more rapidly growing cells, and cell populations retain the same rate-normal distribution and mean cell cycle times over many generations.

These observations led to the presentation of two basic models to explain the variability of cell cycle times. The first is the "deterministic model" proposed in 1962 by Koch and Schaechter.[5] This is based on the assumption that cells initiating their cycle are functionally different, and that cell cycle variability arises from the cumulative effects of many small differences. The second is the "probabilistic model," originally proposed by Burns and Tannock in 1970[6] and later extended by Smith and Martin.[7] This is based on the assumption that cells initiating their cycle are functionally identical, and that the

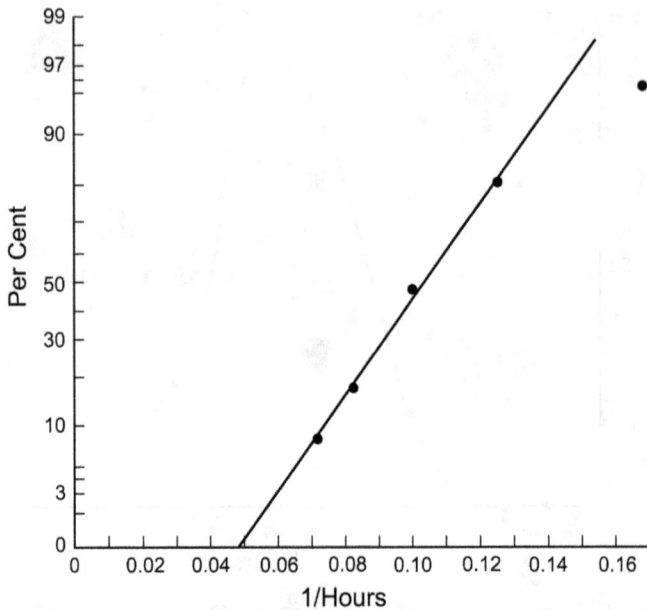

Figure 5.3:   Probit analysis of the rates of each cell traversing G_1. Cumulative distribution of the G_1 "rates" derived by taking the reciprocal of G_1 intervals from Fig. 5.2. (Redrawn from: Sisken, J. and Morasca, L. 1965. *J. Cell Biol.* **25**:179–189.)

transition of cells from an indeterminate quiescent phase to a determinate proliferative phase is regulated by a single Poissonian event, quite independent from other events or properties. Thus, the deterministic model states that there are so many variables between cells that it will be very difficult to ever determine those most important for the variability of cell cycle times, while the probabilistic model states that it will be impossible, because the event is a matter of chance.

In early studies of the growth of cells *in vitro*, it was found that normal mouse embryo fibroblasts (MEF) could be propagated apparently indefinitely if the density of the cells was kept low, so that the cells never contacted one another.[8,9] Thus, MEF cells passed at a density of 3000 cells every three days were termed 3T3 cells, and could be shown to become quiescent if allowed to grow to confluence. This cessation of growth exhibited by cell contact came to be known as "contact inhibition," and was taken as an attribute of normal cells in that 3T3 cells transformed by viruses lost this attribute and could grow to much higher cell densities.

In addition to contact inhibition, 3T3 cells as well as chicken embryo fibroblasts could be made to cease proliferation by removal of serum from the growth media, so that these cells were said to be "serum-dependent."[10] This characteristic also came to be considered as a characteristic of normal cells, in that virus or chemically transformed cells could often continue to proliferate in media with low or absent serum additive.[11] However, the nature of the entities in serum conducive to cell proliferation was not obvious. Thus, it was not clear whether the serum factors were solely nutritional or whether serum components were actually signaling cellular DNA synthesis and mitosis.

Subsequent experiments by Arthur Pardee defined a point late in the G_1 phase of the cell cycle after which removal of serum did not lead to cessation of proliferation.[12] This point, termed the restriction point or "R-point," ultimately became critical for understanding the molecular events responsible after it was realized that sera were not simply nutritional but instead contained molecules that actually signal key events in the cell cycle. Moreover, the length of the G_1 phase was shown to be serum-concentration-dependent.

Pledger and Stiles and their co-workers then performed more detailed experiments, examining the movement of a quiescent population of MEFs to initiate DNA synthesis.[13] They found that the process by which cells leave the quiescent state (i.e. G_0) and enter the G_1 phase of the cell cycle is distinct from the G_1-S-phase transition, and these two transitions are controlled by separate serum components. The first stage, termed "competence," is a prerequisite for entry into the growth cycle, but is insufficient to signal DNA replication. Platelet-derived growth factor (PDGF) is now known to promote MEF cell cycle competence, and the rate of the transition from G_0 to G_1 is PDGF concentration-dependent. The second stage they termed "G_1 progression." They found this to be due to distinct factors found in platelet-poor plasma, now known to be due to insulin-like growth factor-1 (IGF-1) and epidermal growth factor

(EGF). This stage is responsible for movement of the cell beyond the "R-point" in G_1 to the S-phase and subsequent mitosis, and was also found to be dependent on the concentration of the platelet-poor plasma.

By 1980, it was clear that it was important to proceed beyond mathematical analyses of studies of the growth of cells within populations stimulated to grow simply by the addition of serum or nutrients, and to correlate kinetic studies of cell growth with quantitative measurements of the molecules involved in determining competence and progression through the cell cycle. However, to do so required the identification of the critical molecules that promote the cell cycle progression of each individual cell. Because serum was employed as the growth stimulus in the fibroblast experiments, this was a very difficult task.

References

1. Pardee, A., Shilo, B., and Koch, A. (1979) Variability of the cell cycle. In *Hormones and Cell Culture*. G. Sato, and R. Ross, editors. Cold Spring Harbor, NY: Cold Spring Harbor Laboratory. 373–392 pp.
2. Cook, J., and Cook, B. (1962) Effect of nutrients on the variation of individual generation times. *Exp. Cell. Res.* **28**:524–534.
3. Sisken, J., and Morasca, L. (1965) Intrapopulation kinetics of the mitotic cycle. *J. Cell Biol.* **25**:179–189.
4. Jordan, G. (1972) Basis for the probit analysis of an interferon plaque reduction assay. *J. Gen. Virol.* **14**:49–61.
5. Koch, A., and Schaecter, M. (1962) A model for statistics of the cell division process. *J. Gen. Microbiol.* **29**:435–444.
6. Burns, V., and Tannock, I. (1970) On the existence of a G0 phase in the cell cycle. *Cell Tissue Kinetics* **3**:321–333.
7. Smith, J., and Martin, L. (1973) Do cells cycle? *Proc. Natl. Acad. Sci. USA* **70**:1263–1269.
8. Todaro, G., and Green, H. (1963) Quantitative studies of the growth of mouse embryo cells in culture and their development into established cell lines. *J. Cell. Biol.* **17**:299–313.
9. Todaro, G., Lazar, G., and Green, H. (1965) The initiation of cell division in a contact-inhibited mammalian cell line. *J. Cell. Physiol.* **66**:325–333.
10. Temin, H. (1971) Stimulation by serum of multiplication of stationary chicken cells. *J. Cell. Physiol.* **78**:161–170.

11. Temin, H. (1969) Control of cell multiplication in uninfected chicken cells and chicken cells converted by avian sarcoma viruses. *J. Cell. Physiol.* **74**:9–15.
12. Pardee, A. (1974) A restriction point for control of normal animal cell proliferation. *Proc. Natl. Acad. Sci. USA* **71**:1286–1292.
13. Pledger, W., Stiles, C., Antoniades, H., and Scher, C. (1977) Induction of DNA synthesis in BALB/c 3T3 cells by serum components: re-evaluation of the commitment process. *PNAS* **74**:4481–4485.

Chapter 6

The Quantal Nature of IL-2-Promoted T Cell Cycle Progression

To understand how the extremely complex immune system generates a quantal decision as to whether there is immunity to an antigen, reductionism is the only possible experimental approach. In order to apply reductionism to this situation, it is helpful to begin with the end result. This means studying lymphocytes after they have already undergone a proliferative clonal expansion having received the molecular signals initiated by an antigen that result in expression of the mitogenic cytokines and their receptors, as well as the result of the cytokine/receptor interaction, which results in the activation of signaling molecules, TFs and genetic programs that actually drive cell cycle progression and subsequent differentiation to effector function. In this discussion, one premise is that the same principles found to govern the IL-2/IL-2R interaction govern the ligand/receptor interactions of all cytokine/receptor pairs. Moreover, even though the discussion to follow is focused on T cells, another premise is that the principles are similar for B cells and indeed for all cells that make up a metazoan organism, the only differences being the tissue/cell-specific cytokine/receptor pairs, and the signaling pathways, TFs and genes that they activate. Thus, we are discussing fundamental biological processes.

Once the IL-2 molecule had been purified to homogeneity in sufficient quantities so that its concentration could be accurately determined, thereby enabling a specific biological activity to be assigned, the 50% effective concentration (EC_{50}) was found to be quite low, only ~10 pM.[1] Moreover, once the generation and purification of radiolabeled IL-2 was accomplished, we developed the first cytokine binding assay which enabled the quantification of the number and affinity of IL-2 receptors.[2] TCR-activated T cell populations were found to express a mean of ~1000 high affinity IL-2 binding sites/ cell, with an equilibrium dissociation constant (K_d) of 10 pM, thereby indicating that IL-2 biological activity and IL-2 high affinity binding are equivalent, and indicating that there is no rate-limiting molecular step beyond the cell surface IL-2/IL-2R interaction. Also important, resting unstimulated T cells, as well as resting lipopolysaccharide-activated B cells, did not express detectable high affinity IL-2 binding sites. In addition, monoclonal antibodies reactive with both IL-2 and the IL-2R were generated.[1,3] Thus, with the added advantage of monoclonal functional T cells, the cellular and molecular reagents were at hand so that the crucial molecular and cellular parameters responsible for T cell cycle progression could be quantified.

The very first experiments with the IL-2 bioassay using cloned antigen-specific cytolytic T lymphocyte lines (CTLL), revealed the now familiar symmetrically sigmoid log-dose response curve, as shown in Fig. 6.1.[4] In these types of experiments 1 Unit/mL of biological activity was arbitrarily defined as the concentration capable of eliciting an EC_{50} at a dilution of 1:10. Subsequently, 1 Unit/mL was found to be ~10 pM (150 pg/mL), given an IL-2 molecular weight that we determined to be 15.5 KDa.

Although the sigmoid log-dose response curve is familiar to those skilled in pharmacology or toxicology, it was not obvious why the response curve was dependent upon logarithmic IL-2 concentrations, spanning two orders of magnitude. Since these data were generated using cloned T cell populations, one could not ascribe this characteristic heterogeneity to a genetic variation of the cells within the population. Also, since tritiated thymidine (^{3}H-TdR) incorporation into newly synthesized DNA was used to monitor S-phase transition of the cell population, at the EC_{50} it was not obvious whether all of the cells had incorporated only 50% of the maximal ^{3}H-TdR possible, or only half of the cells had entered S-phase. However, it was clear from the radiolabeled IL-2 binding assay that the sigmoid biological dose-response curve and the binding dose-response curves were

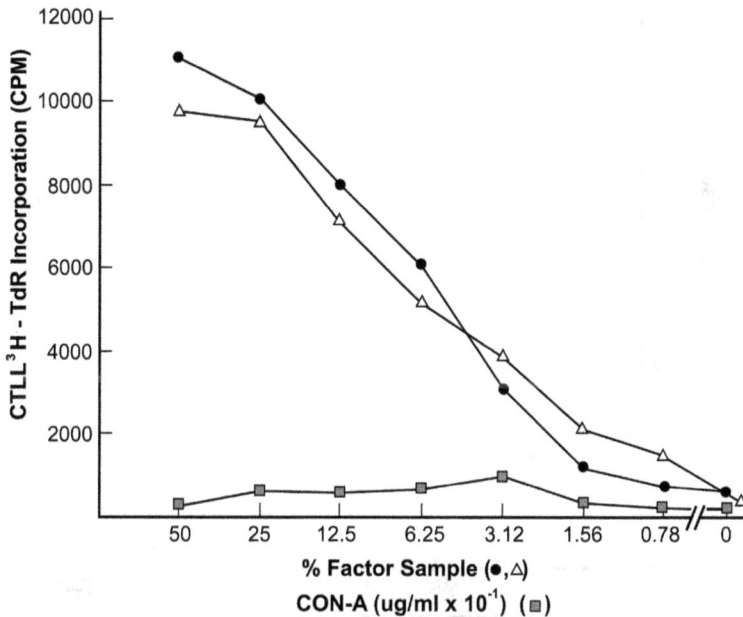

Figure 6.1: Proliferative response of cytotoxic T lymphocyte lines (CTLL) to varying TCGF concentrations. Tritiated thymidine incorporation on the y-axis versus two-fold dilutions of ConA-stimulated Lymphocyte Conditioned Medium (LyCM) (●), LyCM with ConA removed (△), or ConA-containing fresh medium (□) after 24 hours of culture. (Redrawn from: Gillis, S. *et al.* 1978. *J. Immunol.* **120**:2027–2032.)

almost superimposable, as shown in Fig. 6.2.[5] Accordingly, for the cell population, at the $EC_{50} = 10$ pM IL-2, IL-2 receptor binding at steady state was half-maximal and nearly equal to the binding site equilibrium dissociation constant (K_d).

To move beyond cell population studies, it was necessary to be able to monitor each individual cell within the population, and determine whether it was in S-phase or not. Fortunately, Leonard Herzenberg and colleagues had recently introduced the flow cytometer to immunology.[6,7] Thus, using propidium iodide which quantitatively binds DNA, Doreen Cantrell showed that the cells comprising the population enter S-phase in a highly variable fashion when exposed to varying concentrations of IL-2.[8] Thus, some cells

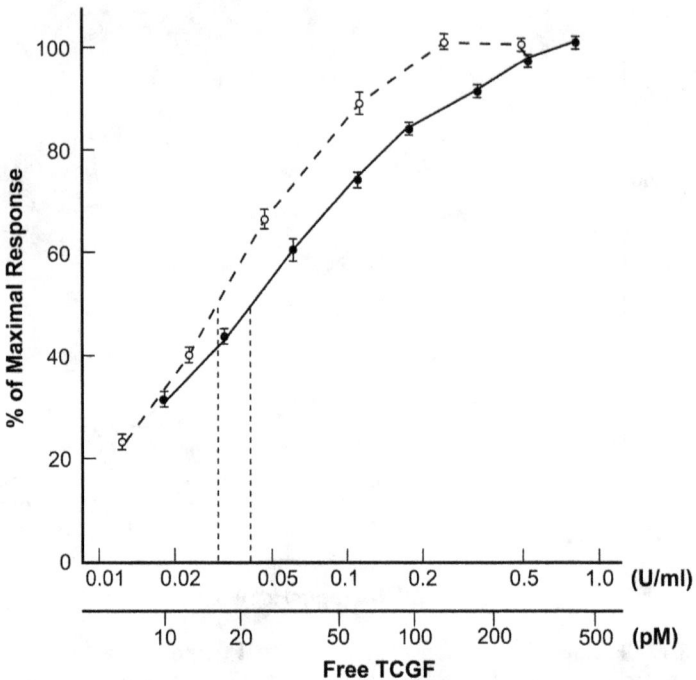

Figure 6.2: Comparison of the TCGF biological dose-response curve with radiolabeled TCGF binding curve. The results are plotted as a percentage of maximum tritiated thymidine incorporation or molecules bound/cell, and the concentrations of TCGF are on the x-axis, either as biological U/mL or as pM purified protein concentrations. (Redrawn from: Robb, R.J. *et al.* 1981. *J. Exp. Med.* **154**:1455–1474.)

enter S-phase after 24 hours of exposure to low concentrations of IL-2, while others require higher concentrations, and at the EC_{50}, half of the cells enter S-phase while half do not (Fig. 6.3).

In addition to demonstrating the variability of responsiveness of the cells within an asynchronously proliferating T cell population, these data also showed that monitoring ^{3}H-TdR incorporation over short time intervals (i.e. one hour) reflects the *proportion* of cells within the population that are in the replicative phase (S-phase) of the cell cycle at that time. The variability in IL-2 responsiveness of an asynchronously proliferating population led to the hypothesis that a

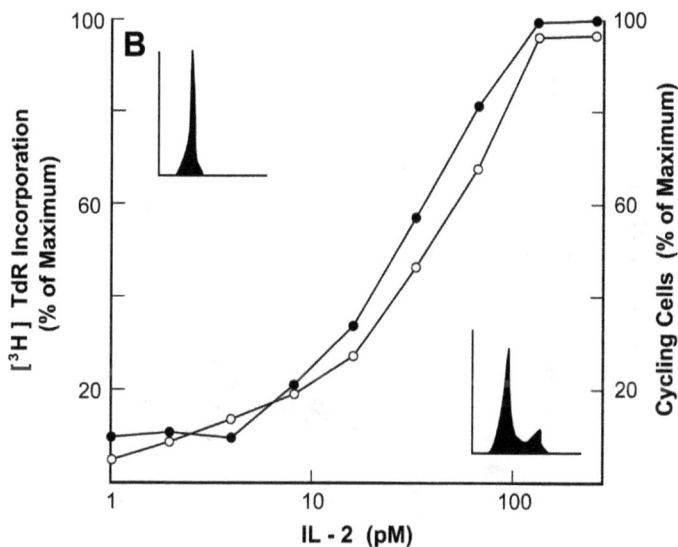

Figure 6.3: **IL-2 proliferative log-dose response curve.** Asynchronously proliferating cells were exposed to varying IL-2 concentrations (x-axis), and tritiated thymidine incorporation by the cell population (●) and Propidium Iodide quantitative DNA determinations by individual cells (○) were monitored after 24 hours of culture. Data are expressed as a % of maximum (y-axis). (Redrawn from: Cantrell, D.A. and Smith, K.A. 1984. *Science* **224**:1312–1318.)

synchronized population of cells would enter the DNA replicative phase of the cell cycle also in a highly variable nature in response to IL-2 with respect to time, just as synchronized MEFs had variable G_1 intervals and cell cycle times in response to the readdition of serum as a growth stimulus. Thus, as shown in Fig. 6.4, ^{3}H-TdR incorporation, and thus the proportion of cells that enter S-phase, gradually and progressively increases over time. Moreover, this variability in the duration of G_1 is also IL-2 concentration-dependent, as depicted below.

The fact that both IL-2 concentration and time determine the proportion of cells entering S-phase indicated that there must be some crucial interval, like the "R-point" of serum-stimulated fibroblasts, when IL-2 was necessary for cell cycle progression, beyond which it would no longer be necessary. From IL-2R kinetic binding

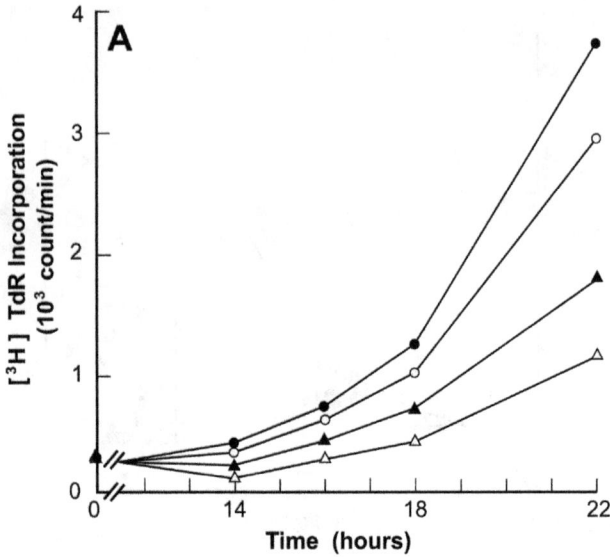

Figure 6.4: IL-2 concentration-dependent cell cycle progression. $G_{0/1}$ Synchronized, IL-2R+ T cells were exposed to varying IL-2 concentrations: 25 pM (△), 50 pM (▲), 100 pM (○), and 500 pM (●). At the indicated intervals, tritiated thymidine incorporation was determined after short (one hour) exposures. (Redrawn from: Cantrell, D.A. and Smith, K.A. 1984. *Science* **224**:1312–1318.)

studies, it had been determined that IL-2 binding reaches a steady state within 10–15 minutes. However, the IL-2/IL-2R interaction had to be sustained for much longer intervals to initiate cell cycle progression, as shown in Fig. 6.5. Thus, when a G_0-G_1 synchronized, IL-2R+ T cell population is exposed to a receptor-saturating IL-2 concentration (250 pM) for varying intervals, a 3-hour exposure is insufficient to trigger any detectable DNA synthesis over the ensuing 26 hours. A minimum of six hours is required for triggering detectable cell cycle progression; moreover, exposure times in excess of six hours result in a progressively greater proportion of cells within the population entering S-phase. In the experiment shown in Fig. 6.5 (inset), exposure times of 6, 11, and 26 hours result in increasingly greater ^{3}H-TdR incorporation at each interval studied, and about 50% of the cells enter S-phase after 11 hours of IL-2 exposure.

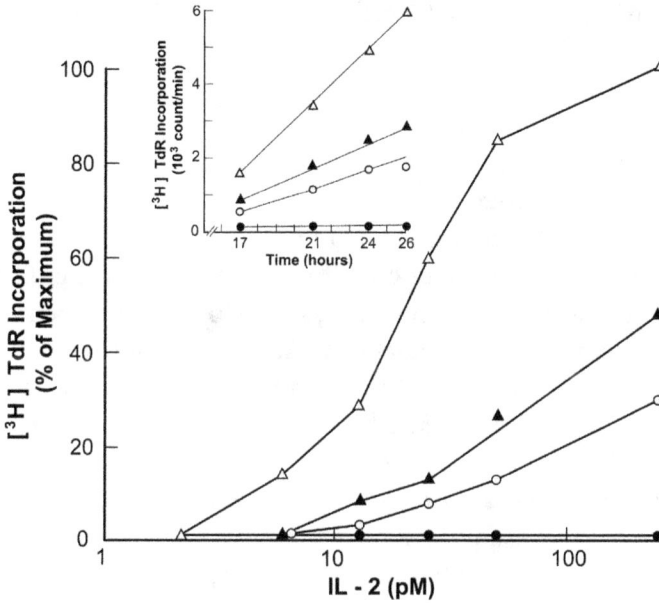

Figure 6.5:   The effect of varying IL-2 exposure time on the cell cycle progression of synchronized G$_{0/1}$ IL-2R$^+$ T cells. The data shown represent the % of maximal tritiated thymidine incorporation (y-axis) plotted as a function of IL-2 concentration (x-axis) after 26 hours of culture. IL-2 exposure times were three hours (●), six hours (○), 11 hours (▲), and 26 hours (△). The inset shows the tritiated thymidine incorporation of each population (i.e. 3, 6, 11, and 26 hours) at an IL-2R saturating concentration (250 pM) monitored at the times indicated. (Redrawn from: Cantrell, D.A. and Smith, K.A. 1984. *Science* **224**:1312–1318.)

Accordingly, both IL-2 concentration and the duration of the IL-2 cellular interaction are critical determinants of cell cycle progression. There appears to be interplay between these two variables, such that the *proportion* of cells responsive to suboptimal IL-2 concentrations can be increased by lengthening the exposure period.

When whole serum had been used as the growth stimulus for MEFs, it was impossible to go beyond these kinds of experiments to focus on the critical variables determining cell cycle progression. However, using monoclonal T cells, purified homogeneous IL-2, MoAbs reactive with both IL-2 and the IL-2R, together with the capacity to monitor single cell cycle progression using the flow cytometer, new approaches to the variables regulating cell cycle progression were possible for the first time. Since conditions that limited the IL-2-cellular interaction modified the *proportion* of cells that progressed to S-phase, the conclusion was inescapable that there must be an underlying heterogeneity among the cells comprising the population with regard to the capacity to respond to a given concentration of IL-2. Such heterogeneity could not be due to genetic differences between cells within the population, since cloned T cells behaved similarly to uncloned cell populations. Also, variability within separate phases of the cell cycle could not be responsible, since synchronized cells behaved similarly to asynchronized cell populations.

Once IL-2Rs could be monitored and it was possible using the flow cytometer to examine individual cells in T cell populations for IL-2R expression using MoAbs reactive with the IL-2R, a semi-log plot of the fluorescent intensity of IL-2R expression versus cell number revealed what is now familiar to all immunologists. There is a log-normal distribution of IL-2Rs on individual cells comprising a population that spans at least two orders of magnitude (Fig. 6.6). Thus, some cells have very few IL-2Rs, while others have 100-fold greater numbers of receptors based upon fluorescent intensity of anti-IL-2R reactivity, with most cells within the population distributed about the mean in a log-normal fashion between the two extremes. The log-normal distribution of IL-2R densities on individual cells within a population was striking when recalling the rate-normal distribution of individual cell cycle times (see Fig. 5.2). Because of the similarities of these plots and because of the importance of the IL-2/IL-2R interaction for driving T cell cycle progression, it was logical that cell cycle time variability was determined by the intrapopulation heterogeneity of IL-2R density per cell. In other words, given a receptor-saturating IL-2 concentration, cells with the highest density of IL-2Rs would be expected to traverse the cell cycle more rapidly than cells with a lower IL-R density.

Cell sorting experiments permitted a direct approach to the analysis of the relevance of IL-2R density to cell cycle progression. As shown in Fig. 6.7, a synchronized IL-2R+ cell population was

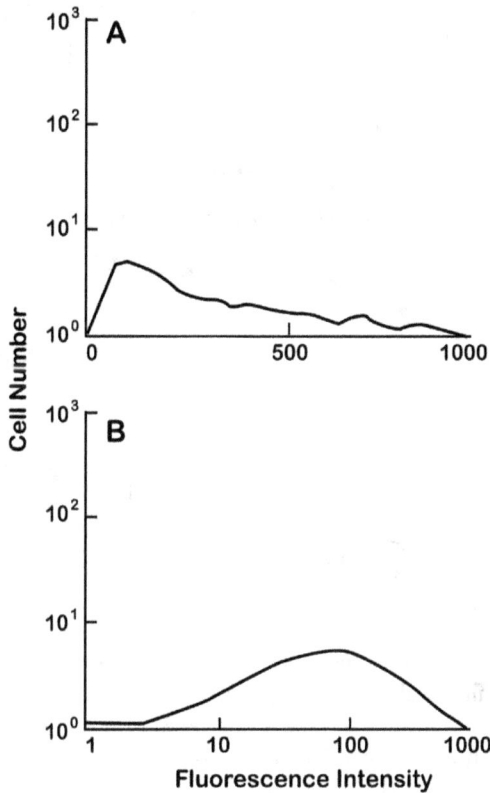

Figure 6.6: **Flow cytometric analysis of anti-Tac binding to human T cells three days after activation with PHA. A.** Linear plot of cell number versus fluorescence intensity. **B.** Logarithmic plot of fluorescent intensity. (Redrawn from: Cantrell, D.A. and Smith, K.A. 1984. *Science* **224**:1312–1318.)

separated into low and high IL-2R density subsets (Figs. 6.7A–6.7C), and exposed to a saturating concentration of IL-2 for varying time intervals before analysis with short ^{3}H-TdR pulses to monitor S-phase transition (Fig. 6.7D). Thus, the high-IL-2R-density subset entered S-phase long before the low-IL-R-density subset, and furthermore, the high-density subset appeared to account for most of the ^{3}H-TdR incorporation of the unseparated cell population. Also, as shown in Fig. 6.6E, as anticipated from experiments with unseparated cell

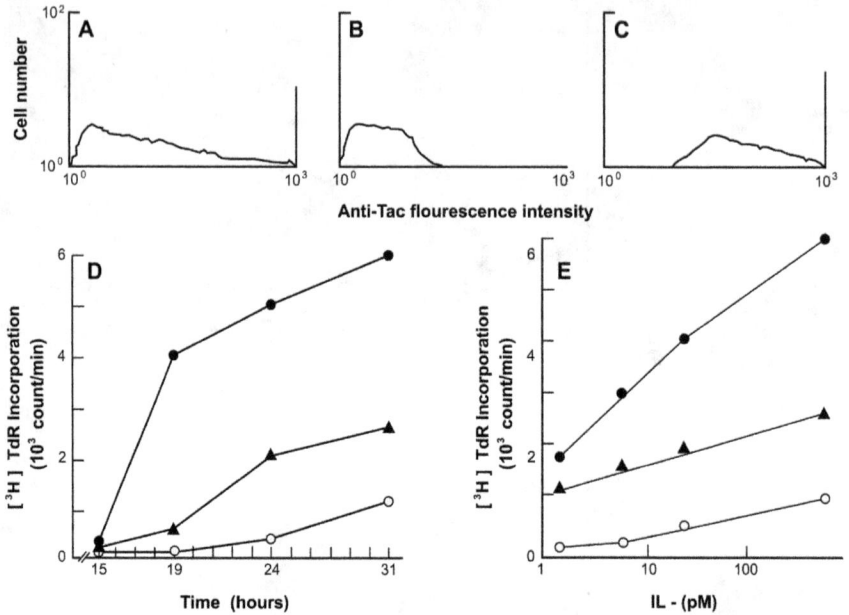

Figure 6.7:   The proliferative response of IL-2R⁺ cell populations separated on the basis of IL-2Rα chain density. A. Linear plot of the IL-2Rα chain density of the unseparated cell population. **B.** Low IL-2Rα chain density subset. **C.** High IL-2Rα chain density subset. **D.** Tritiated thymidine incorporation in response to an IL-2R saturating concentration of IL-2 (250 pM) of unseparated cells (▲), low IL-2Rα chain density subset (O), and high IL-2Rα chain density subset (●). **E.** Tritiated thymidine incorporation of same subsets plotted as a function of IL-2 concentration after 31 hours of culture. (Redrawn from: Cantrell, D.A. and Smith, K.A. 1984. *Science* **224**:1312–1318.)

populations, the response of cell subsets separated on the basis of IL-2R density was dependent upon IL-2 concentration. At limiting IL-2 concentrations, only the high-IL-2R-density subset could be detected in the S-phase of the cell cycle. Thus, the log-normal distribution of cellular IL-2R density of individual cells within a T cell population is responsible for the normal sigmoid shape of the IL-2 log-dose response curve. That is, the cells that respond to IL-2 first are the ones with the highest density of IL-2Rs, while the cells that take the longest to respond are those with the lowest density of IL-2Rs.

Accordingly, there are only four variables regulating T cell cycle progression:

1) the affinity of the IL-2/IL-2R interaction
2) the IL-2 concentrations necessary, determined by the first variable given above
3) the density of IL-2Rs/cell
4) the duration that IL-2 and IL-2Rs interact.[8]

This last variable, time, can be lengthened or shortened by raising or lowering either the IL-2 concentration or the IL-2R density. Because each cell progresses beyond the R-point in a quantal fashion, and the variables (molecules) determining this G_0-G_1 transition are now known, the only possible interpretation is that there is some crucial, definite number of IL-2/IL-2R interactions that must be satisfied. Moreover, the cell appears to somehow be able to "count" these IL-2/IL-2R interactions, and wait until the crucial number has accumulated.

It follows that if this crucial number of IL-2/IL-2R interactions is not met, then despite antigen activation of the TCR, there will be no proliferative expansion of this clone of T cells, and consequently no detectable response at the systemic level to the antigen activation event. In other words, ultimately the control of responsiveness to the introduction of antigenic molecules from the environment is undertaken by an endogenous hormone/receptor system.

References

1. Smith, K.A., Favata, M.F., and Oroszlan, S. (1983) Production and characterization of monoclonal antibodies to human interleukin 2: strategy and tactics. *J. Immunol.* **131**:1808–1815.
2. Robb, R.J., Munck, A., and Smith, K.A. (1981) T cell growth factor receptors: quantitation, specificity, and biological relevance. *J. Exp. Med.* **154**:1455–1474.
3. Leonard, W.J., Depper, J.M., Uchiyama, T., Smith, K.A., Waldmann, T.A., and Greene, W.C. (1982) A monoclonal antibody that appears to recognize the receptor for human T-cell growth factor; partial characterization of the receptor. *Nature* **300**:267–269.

4. Gillis, S., Ferm, M.M., Ou, W., and Smith, K.A. (1978) T cell growth factor: parameters of production and a quantitative microassay for activity. *J. Immunol.* **120**:2027–2032.
5. Robb, R.J., and Smith, K.A. (1981) Heterogeneity of human T-cell growth factor(s) due to variable glycosylation. *Mol. Immunol.* **18**:1087–1094.
6. Julius, M., Masuda, T., and Herzenberg, L. (1972) Demonstration that antigen-binding cells are precursors of antibody-producing cells after purification with a fluorescence-activated cell sorter. *Proc. Natl. Acad. Sci. USA* **69**:1934–1938.
7. Parks, D.R., Bryan, V.M., Oi, V.T., and Herzenberg, L.A. (1979) Antigen-specific identification and cloning of hybridomas with a fluorescence-activated cell sorter. *Proc. Natl. Acad. Sci. USA* **76**:1962–1966.
8. Cantrell, D.A., and Smith, K.A. (1984) The interleukin-2 T-cell system: a new cell growth model. *Science* **224**:1312–1316.

Chapter 7

The Molecular Basis for Quantal IL-2/IL-2R Signaling of Cell Cycle Progression — The IL-2/Receptor Interaction

From quantitative IL-2/IL-2R equilibrium and kinetic binding experiments, it was found that there is a mean of ~1000 high affinity IL-2R sites/cell on normal human antigen-activated T cells that bind IL-2 with a very rapid association rate ($\kappa = 1 \times 10^7$ sec^{-1}M^{-1}) and a relatively slow dissociation rate ($\kappa' = 1 \times 10^{-4}$ sec^{-1}).[1] Consequently, steady state binding of IL-2 occurs within 10–15 minutes. However, as already noted, synchronized T cells require several hours of IL-2 exposure to lead to the quantal decision to progress beyond the R-point, synthesize nucleotides and replicate DNA. Thus, occupancy of only 1000 IL-2R sites for 10 minutes does not lead to the intracellular changes necessary for G$_1$ progression. Just like serum-stimulated MEFs, if IL-2 is withdrawn before the R-point, the cell will not make the G$_1$-S-phase transition. For the first time the structures of the IL-2 molecule and the IL-2R chain molecules could be examined to determine how the IL-2/IL-2R interaction occurs and initiates the signaling process in T cells, increasing our understanding of how this quantal decision is made. It must be noted from the start that to signal a quantal cellular proliferative response the molecular processes must be "hard-wired" so that the quantitative information of the number of IL-2/IL-2R interactions that occur at the cell surface is transferred to the cell interior, and ultimately to the nucleus.

IL-2 was the first cytokine to be examined by X-ray crystallography. An initial structure resolved at 3 Å in 1987 revealed that there are four alpha helices,[2] which was consistent with earlier data obtained by circular dichroism.[3] Subsequent comparisons with the structures of GM-CSF and IL-4 led to refinement of the IL-2 structure, which at 2.5 Å resolution five years later showed that it has what is now called the common cytokine fold, with typical up-up (helices A and B) down-down (helices C and D) four-helix topology, as shown in Fig. 7.1.[4]

Over the course of a decade, three non-covalently linked chains (termed α, β, and γ) were found to comprise the high affinity IL-2R.[5-9] Since the IL-2R is made up of three non-covalently linked chains, it is necessary to understand how the physical characteristics of IL-2/IL-2R binding, i.e. the rapid on-rate and slow off-rate are created. The first such studies by Huey-Mei Wang showed that the rapid on-rate of IL-2 binding to the heterotrimeric IL-2R is contributed by

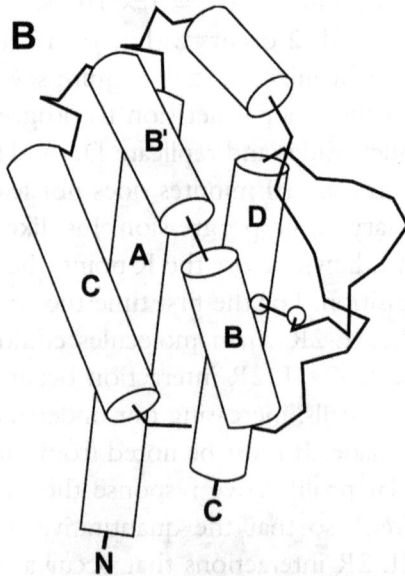

Figure 7.1:   Schematic drawing of [Ala125] IL-2. The α helices are shown as cylinders. (Redrawn from: Bazan, J.F. and McKay, D.B. 1992. *Science* **257**:410–413.)

the IL-2Rα chain, while the slow off-rate is contributed by the IL-2Rβ chain.[10] Subsequently, Thomas Ciardelli's group used surface plasmon resonance (SPR) and purified IL-2R proteins, and found IL-2 to bind to isolated IL-2Rα chains with a rapid on-rate ($\kappa = 1.1 \times 10^6$ M^{-1}s^{-1}), but it also dissociates rapidly from isolated IL-2Rα chains ($\kappa' = 3.5 \times 10^{-2}$ s^{-1}).[11] Therefore, with a fast on-rate and fast off-rate, the K_d (κ'/κ) is modest (~ 30 nM). Such a relatively low affinity interaction necessitates a similar relatively high IL-2 concentration to bind, and at least 300 nM to fully saturate. Since the K_d of the heterotrimeric IL-2R is only 10 pM, a 30,000-fold difference, it is doubtful that such high IL-2 concentrations actually exist physiologically, even within a microenvironment.

The crystallization of IL-2 bound to IL-2Rα chains revealed that the four-helix bundle of IL-2 binds to the IL-2Rα chain via the IL-2 helices A' and B' and part of the AB loop, located at the top of the structure shown in Fig. 7.1.[12] As a result of complex formation with the IL-2Rα chain, there is a total buried surface area of 1868 Å^2 between IL-2 and the IL-2Rα chain, with a hydrophobic center and a polar periphery featuring five ion pairs and six hydrogen bonds. There are two prominent hydrophobic patches on IL-2 in the center of the IL-2 site, which is consistent with thermodynamic measurements indicating that the desolvation of an apolar surface is the primary driving force of this interaction.

The earliest studies revealed ~100-fold greater number of low affinity IL-2Rα chains than high affinity binding sites.[13] Thus, if the IL-2 concentrations are high enough to saturate the IL-2Rα chains, then IL-2 binding to isolated IL-2Rα chains could serve to change the search of IL-2 for IL-2Rs from a three-dimensional search in space to a two-dimensional search in the plane of the membrane, in that once bound to IL-2Rα chains, the formation of a quaternary complex of IL-2 bound to all three IL-2R chains could proceed more readily in the plane of the membrane.

Alternatively, by surface plasmon resonance (SPR) experiments using purified IL-2R chains[14] as well as by isothermal calorimetry (ITC),[15] it has been determined that IL-2Rα chains can bind to IL-2Rβ chains (K_d = 278 nM) in the absence of IL-2, so that the

100-fold excess of IL-2Rα versus IL-2Rβ chains would favor the formation of α/β heterodimers on the cell surface by the law of mass action. Also, no detectable binding between IL-2Rα and IL-2Rγ chains or between IL-2Rβ chains and IL-2Rγ chains occurs in the absence of IL-2, thereby indicating that both IL-2Rβ chains and IL-2Rγ chains exist on the cell surface independently of one another. IL-2 could be captured by an α/β heterodimer more efficiently than by IL-2Rα chains alone, because IL-2 has a ~2 to 3-fold more rapid on-rate to this α/β heterodimer ($\kappa = 2.3 \times 10^6$ M^{-1}s^{-1}) compared with isolated IL-2Rα chains.[16] Even more compelling, IL-2 has a much slower off-rate (>10-fold) from the α/β heterodimer ($\kappa' = 0.018$ s^{-1}), thereby leading to a higher binding affinity (K$_d$ = 300 pM), which has been termed pseudo-high affinity binding. By comparison, even though IL-2 can bind to isolated IL-2Rβ chains, the on-rate is relatively slow, which combined with a slow off-rate, makes for a comparatively very low affinity interaction (K$_d$ = 150 nM). Finally, IL-2 can only very weakly interact with isolated IL-2Rγ chains (K$_d$ > 700 μM), which is consistent with the observations that the IL-2Rγ chain is also called the common γ chain (γ_c), because it serves as a part of the receptors for several other cytokines, including IL-4, IL-7, IL-9, IL-15, and IL-21.[17]

It has also been determined that signaling does not occur until the γ_c chain is recruited to join the IL-2/IL-2Rα,β trimeric complex, which is consistent with the notion that both β and γ_c chains participate in signaling the cell interior. Even though the γ_c chain is only weakly able to interact with IL-2 by itself, the participation of the γ_c chain reduces the off-rate of the heterodimer-bound ligand substantially by forming a very stable quaternary ligand/receptor complex.[18] The SPR and ITC experiments revealed that the trimer formed by IL-2 bound to α and β chains is able to bind to the γ_c chain (K$_d$ = 12 nM). From the crystallization studies of this quaternary complex by both Christopher Garcia's[19] and Ian Wilson's[20] groups it was found that the sites on IL-2 that interact with the three chains do not overlap. The opposite side of the IL-2 four-helix bundle held by the α-chain is clamped between the elbow regions of the β and γ chains, which converge to form a Y shape, with IL-2 bound in the fork of the Y.

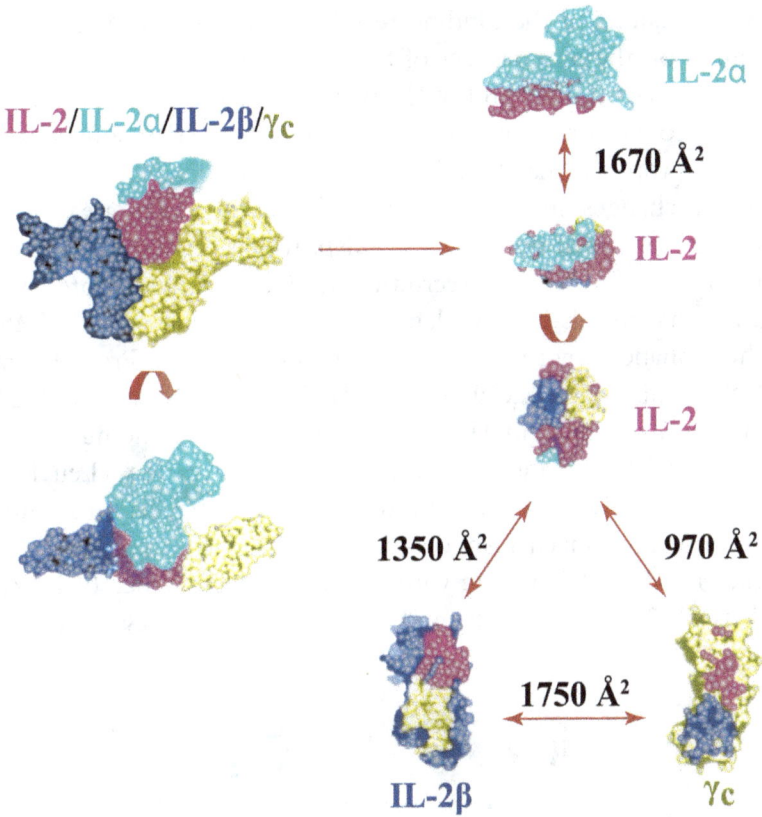

Figure 7.2: **"Open Book" representations of surface interactions between the molecules of the IL-2/IL-2R quaternary complex.** Each color is shown in the contacted area of the cytokine or the receptor chains. (From: Wang, X. *et al.* 2005. *Science* **310**:1159–1163.)

An "open book" surface representation of the binding interfaces within the quaternary ligand/receptor complex is shown in Fig. 7.2. It can be readily appreciated that IL-2Rα binds a face of IL-2 distinct from that bound by IL-2Rβ and IL-2Rγ_c. Moreover, there is an extensive interaction face between IL-2Rβ and IL-2Rγ_c, with IL-2 held between these two chains.

However, knowing the structures of these molecules and where they interact does not lead to any conclusions as to *how* they interact, i.e. the dynamics of the binding reactions and how this is translated to signals on the inner aspect of the membrane. Because the buried surface areas of IL-2Rβ and of IL-2Rγ_c that are involved in receptor/receptor contacts and receptor/IL-2 contacts overlap somewhat, it has been proposed that IL-2 binding to IL-2Rβ may induce conformational changes in its contiguous IL-2Rγ_c contact area of the membrane proximal portion of IL-2Rβ, which primes the IL-2/IL-2Rα/IL-2Rβ for IL-2Rγ_c recruitment. It can be appreciated from Fig. 7.3 that the IL-2-dependent IL-2Rβ association with IL-2Rγ_c is further enhanced by three IL-2Rγ_c residues (Tyr-182$^\gamma$, Pro-207$^\gamma$, and L-208$^\gamma$), which interface with both IL-2 and IL-2Rβ, two IL-2Rβ residues (Glu-136$^\beta$, and His-138$^\beta$) that are correspondingly buried in the IL-2/IL-2Rγ_c interface and by three IL-2 residues (Leu-12^{IL-2}, Glu-15^{IL-2}, and Leu-19^{IL-2}) that interface with both receptor subunits. This cooperative mechanism could then account for the undetectable affinities of IL-2Rβ binding with IL-2Rγ_c in the absence of IL-2, as well as IL-2 with IL-2Rγ_c, and would provide an essential safety

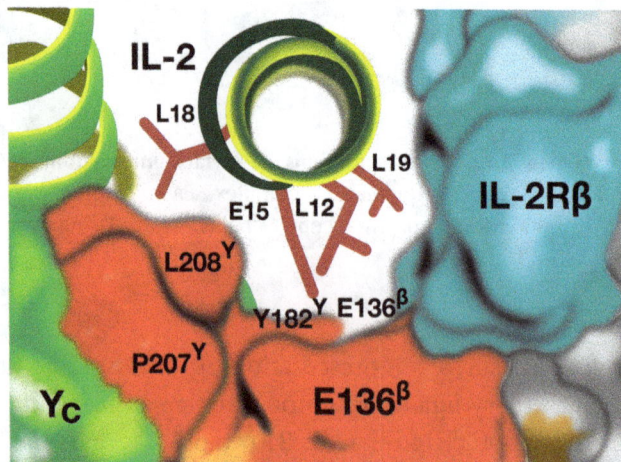

Figure 7.3:   Three-way junction between IL-2, IL-2Rβ, and IL-2Rγ_c. IL-2, IL-2Rβ, and IL-2Rγ_c form a three-way junction at the heart of the high affinity IL-2 signaling complex. The network of residues that mediate these contacts provides a compelling structural basis for cooperativity in the IL-2/IL-2R complex assembly. (Redrawn from: Stauber, D. *et al.* 2006. *PNAS* **103**:2788–2793.)

mechanism against premature signaling via an IL-2Rβ/IL-2Rγ_c dimer, if it could form in the absence of IL-2.

All of this biophysical and structural data are entirely consistent with what is known about IL-2 binding and IL-2R signaling, in that even in the absence of IL-2 the cytoplasmic domain of the β chain is already complexed with the Janus protein tyrosine kinase (PTK)-1 (JAK-1), while the γ_c chain cytoplasmic domain is already complexed with the JAK-3 PTK.[21,22] Thus, only when IL-2 binding brings the external domains of these two receptor chains together, thereby bringing the cytoplasmic domains into close enough proximity, can signaling occur by trans-phosphorylation, not only of the two PTKs, but of the cytoplasmic domains of the IL-2Rβ/IL-2Rγ_c receptor chains. These latter phosphorylation events of the receptor chains are particularly important, because these sites serve as docking sites for the next molecules in the signaling pathways that transfer the signals to the cell interior. Unfortunately, at this time all efforts to crystallize the cytoplasmic domains of the IL-2R chains have been fruitless, which thus far have been taken as evidence that the cytoplasmic tails do not adopt a permanent well-defined structure, which is a prerequisite for crystallization. However, all evidence points to the formation of a very stable quaternary IL-2/IL-2R signaling complex, which continues to signal until it is internalized and degraded. The regulation of the duration that this stable ligand/receptor complex can signal is thus dependent upon the continued presence and concentration of IL-2, and the rate of internalization and removal of the ligand/receptor complex, together with the rate of synthesis of new receptors.

References

1. Robb, R.J., Munck, A., and Smith, K.A. (1981) T cell growth factor receptors: quantitation, specificity, and biological relevance. *J. Exp. Med.* **154**:1455–1474.
2. Brandhuber, B.J., Boone, T., Kenney, W.C., and McKay, D.B. (1987) Three-dimensional structure of interleukin 2. *Science* **238**:1707–1709.
3. Cohen, F., Kosen, P., Kuntz, I., Epstein, L., Ciardelli, T., and Smith, K. (1986) Structure-activity studies of interleukin-2. *Science* **234**:349–352.
4. Bazan, J.F. (1992) Unraveling the structure of IL2. *Science* **257**:410–413.

5. Leonard, W.J., Depper, J.M., Uchiyama, T., Smith, K.A., Waldmann, T.A., and Greene, W.C. (1982) A monoclonal antibody that appears to recognize the receptor for human T-cell growth factor; partial characterization of the receptor. *Nature* **300**:267–269.

6. Sharon, M., Klausner, R.D., Cullen, B.R., Chizzonite, R., and Leonard, W.J. (1986) Novel interleukin-2 receptor subunit detected by cross-linking under high affinity conditions. *Science* **234**:859–863.

7. Tsudo, M., Goldman, C.K., Bongiovanni, K.F., Chan, W.C., Winton, E.F., Yagita, M., Grimm, E.A., and Waldmann, T.A. (1987) The p75 peptide is the receptor for interleukin 2 expressed on large granular lymphocytes and is responsible for the interleukin 2 activation of these cells. *Proc. Natl. Acad. Sci. USA* **84**:5394–5398.

8. Teshigawara, K., Wang, H.M., Kato, K., and Smith, K.A. (1987) Interleukin 2 high-affinity receptor expression requires two distinct binding proteins. *J. Exp. Med.* **165**:223–238.

9. Takeshita, T., Ohtani, K., Asao, H., Kumaki, S., Nakamura, M., and Sugamura, K. (1992) An associated molecule, p64, with IL-2 receptor beta chain. Its possible involvement in the formation of the functional intermediate- affinity IL-2 receptor complex. *J. Immunol.* **148**:2154–2158.

10. Wang, H.M., and Smith, K.A. (1987) The interleukin 2 receptor. Functional consequences of its bimolecular structure. *J. Exp. Med.* **166**:1055–1069.

11. Wu, Z., Johnson, K.W., Goldstein, B., Choi, Y., Eaton, S.F., Laue, T.M., and Ciardelli, T.L. (1995) Solution assembly of a soluble, heteromeric, high affinity interleukin-2 receptor complex. *J. Biol. Chem.* **270**:16039–16044.

12. Rickert, M., Wang, X., Boulanger, M., Goriatcheva, N., and Garcia, K. (2005) The structure of interleukin-2 complexed with its alpha receptor. *Science* **308**:1477–1480.

13. Smith, K.A., and Cantrell, D.A. (1985) Interleukin 2 regulates its own receptors. *Proc. Natl. Acad. Sci. USA* **82**:864–868.

14. Wu, Z., Goldstein, B., Laue, T.M., Liparoto, S., Nemeth, M., and Ciardelli, T. (1999) Solution assembly of the pseudo-high affinity and intermediate affinity interleukin-2 receptor complexes. *Protein Sci.* **8**:482–489.

15. Rickert, M., Boulanger, M., Goriatcheva, N., and Garcia, K. (2004) Compensatory energetic mechanisms mediating the assembly of signaling complexes between interleukin 2 and its alpha, beta and gamma receptors. *J. Mol. Biol.* **339**:1115–1128.

16. Liparoto, S., and Ciardelli, T. (1999) Biosensor analysis of the interleukin-2 receptor complex. *J. Mol. Rec.* **12**:316–321.

17. Leonard, W., Shores, E., and Love, P. (1995) Role of the common cytokine receptor gamma chain in cytokine signaling and lymphoid development. *Immunol. Rev.* **148**:97–114.

18. Liparoto, S.F., Myszka, D.G., Wu, Z., Goldstein, B., Laue, T.M., and Ciardelli, T.L. (2002) Analysis of the role of the interleukin-2 receptor gamma chain in ligand binding. *Biochemistry* **41**:2543–2551.

19. Wang, X., Rickert, M., and Garcia, K.C. (2005) Structure of the quaternary complex of interleukin-2 with its {alpha}, {beta}, and {gamma}c receptors. *Science* **310**:1159–1163.

20. Stauber, D., Debler, E., Horton, P., Smith, K., and Wilson, I. (2006) Crystal structure of the interleukin-2 signaling complex: paradigm for a heterotrimeric cytokine receptor. *PNAS* **103**:2788–2793.

21. Russell, S., Johnston, J., Noguchi, M., Kawamura, M., Witthuhn, B., Silvennoinen, O., Goldman, A., Schmalsteig, F., Ihle, J., O'Shea, J., *et al.* (1994) Interaction of IL2R beta and gamma-c chains with JAK1 and JAK3: implications for XSCID and XCID. *Science* **266**:1042–1045.

22. Miyazaki, T., Kawahara, A., Fujii, H., Nakagawa, Y., Minami, Y., Liu, Z.J., Oishi, I., Silvennoinen, O., Witthuhn, B.A., Ihle, J.N., *et al.* (1994) Functional activation of Jak1 and Jak3 by selective association with IL-2 receptor subunits. *Science* **266**:1045–1047.

Chapter 8

The Molecular Basis for Quantal IL-2/IL-2R Signaling of Cell Cycle Progression — IL-2 and IL-2 Receptor Metabolism

The earliest experiments with IL-2-dependent cloned T cells revealed that as the cells proliferate the IL-2 concentration progressively declines, as shown in Fig. 8.1. Moreover, once IL-2 concentrations decline to low levels, IL-2-dependent cells rapidly undergo apoptosis. Therefore, it appeared that IL-2-dependent cells somehow consume IL-2 and that IL-2 is necessary not only for proliferation but also for survival. As soon as radiolabeled IL-2 was available, IL-2 was found to be degraded into a trichloroacetic acid (TCA) soluble form by IL-2-dependent cells, indicating that the cells metabolize the radiolabeled IL-2 molecules into smaller forms that could not be precipitated by TCA.[1]

The turnover of membrane proteins, which can be determined by ascertaining the half-times $(t_{1/2})$ for the disappearance and appearance of molecules on the cell surface, is equivalent to their median functional lifetime. Therefore, to gain insight into the physiology of IL-2Rs, the turnover characteristics of high affinity IL-2 binding sites were studied and compared with IL-2Rα chain binding sites detected by the anti-Tac MoAb.[2] Using normal human T cells that had been PHA-activated to express maximal densities of both high affinity IL-2Rs as well as low affinity IL-2Rα chains, as shown in Fig. 8.2A, the decay of cell surface binding sites was studied after blocking protein synthesis with cycloheximide. In the absence of IL-2, the $t_{1/2}$ for the

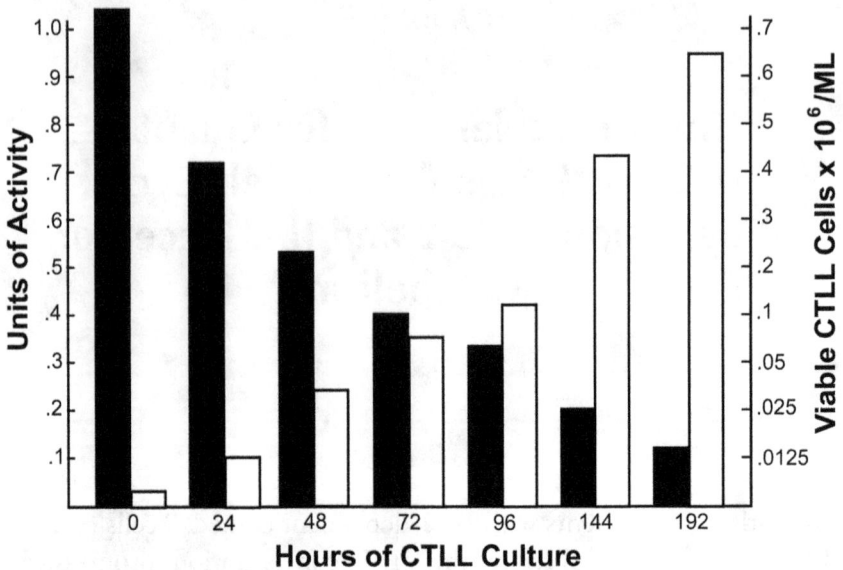

Figure 8.1:   The disappearance of IL-2 (TCGF) as CTLL proliferate. CTLL were seeded at a low concentration (6×10^3 cells/mL) at t = 0 in 1.0 U/mL (~10 pM) IL-2 (TCGF). Daily the supernatants were assayed for IL-2 activity and cell concentrations determined. After eight days (192 hours) of culture the cell density had increased 100-fold and the IL-2 (TCGF) concentration had decreased to very low levels. (Redrawn from: Gillis, S. *et al.* 1978. *J. Immunol.* **120**:2027–2032.)

decay of high affinity IL-2 binding sites was ~ 2.5 hours (150 minutes), while the low affinity IL-2Rα chains had not reached the 50% point even after five hours. The converse experiment, tracking the appearance of new IL-2 binding sites subsequent to limited membrane proteolysis with pronase treatment, is shown in Fig. 8.2B. High affinity IL-2 binding sites returned to 50% of control levels within 2.2 hours and attained 90% of the original levels within five hours, whereas the reappearance of low affinity IL-2Rα chains was remarkably retarded, reaching only ~ 20% of pretreatment levels during this interval.

These experiments indicated that the high affinity IL-2 binding sites and the low affinity IL-2 binding sites were somehow regulated differently. This difference was highlighted when IL-2 was added.

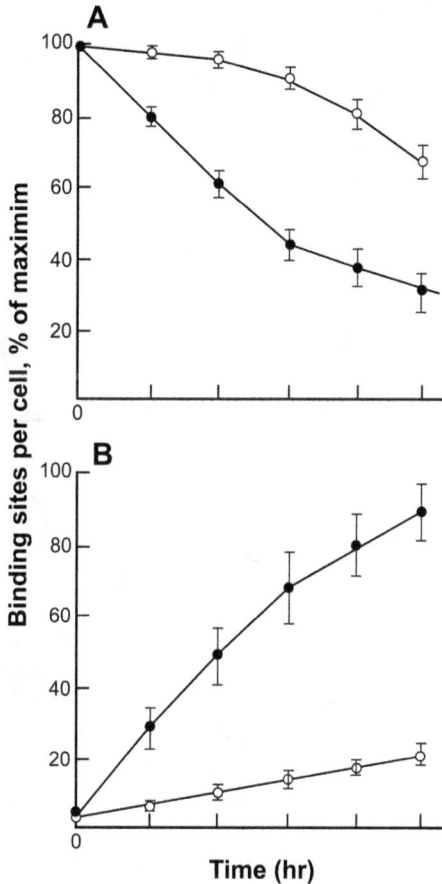

Figure 8.2: The disappearance (A) and reappearance (B) of high affinity IL-2 binding sites (●) and IL-2Rα chains (anti-Tac) (O). In *A*, IL-2R+ cells were exposed to Cycloheximide to block new protein synthesis, thereby monitoring the decay of surface sites over time. In *B*, surface membrane binding sites were stripped by a limited digestion with Pronase, then the reappearance of binding sites was monitored over time. (Redrawn from: Smith, K.A. and Cantrell, D.A. 1985. *PNAS* **82**:864–869.)

As shown in Fig. 8.3, the exposure of G_0/G_1 synchronized IL-2R+ cells to IL-2 resulted in a progressive increase in the density of low affinity IL-2Rα chains over 24 hours, while the density of high affinity IL-2 binding sites decreased by 50% (Inset A), and the density of

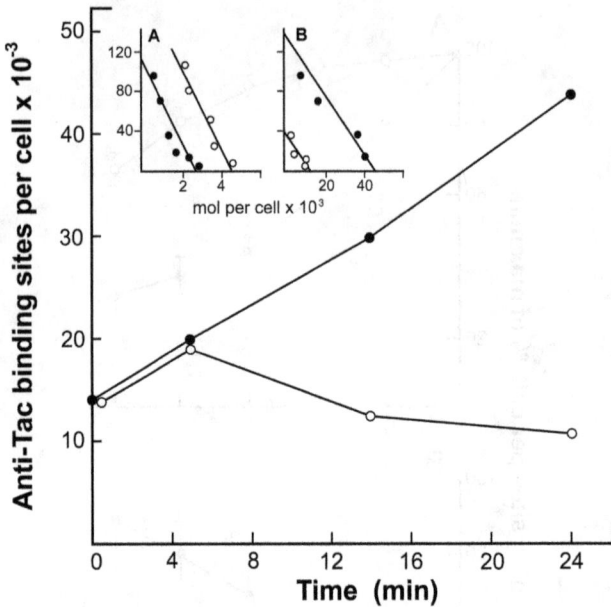

Figure 8.3:   IL-2-mediated augmentation of IL-2Rα chain expression and diminishment of high affinity IL-2R expression. IL-2R+ cells were cultured without IL-2 for 24 hours, then exposed (●) to an IL-2R-saturating IL-2 concentration (500 pM) or not (O). Aliquots of cells were taken at the indicated times and assayed for high affinity IL-2R expression or for IL-2Rα chain expression via anti-Tac. **Inset A:** Scatchard plot of radiolabeled IL-2 binding after 24 hours of exposure to IL-2 (●), or not (O). **Inset B:** Scatchard plot of radiolabeled anti-Tac binding after 24 hours of exposure to IL-2 (●), or not (O). (Redrawn from: Smith, K.A. and Cantrell, D.A. *1985. PNAS* **82**:864–869.)

low affinity IL-2 binding sites increased by >50% over the same time interval (Inset B).

The heterotrimeric structure of high affinity IL-2 binding sites was not fully understood until a decade after the IL-2R had been first described and characterized, and was dependent upon the discovery of the α, β, and γc chains. During this time, and over the course of the next decade, the laboratory of Alice Dautry-Varsat detailed the metabolism of IL-2 and the three IL-2 receptor chains, and ultimately established a new pathway whereby the IL-2R chains are metabolized by the cell.[3-8] As shown in Fig. 8.4A, the disappearance of

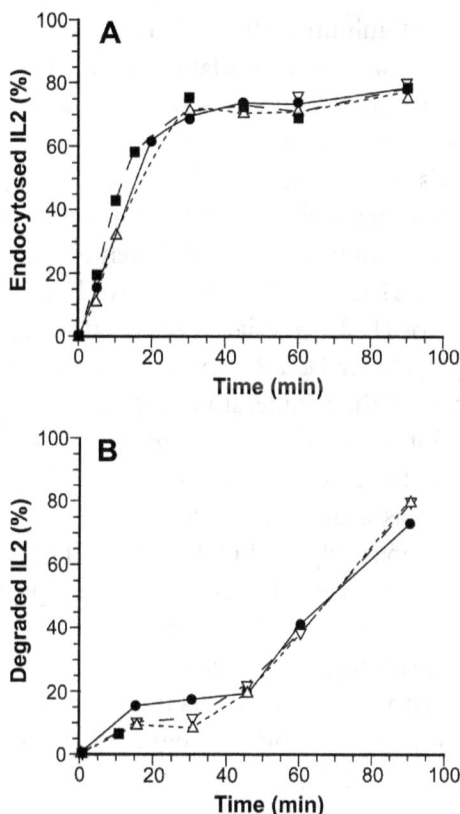

Figure 8.4: **Endocytosis (A) and degradation (B) of** ^{125}**I-IL-2 in the presence and absence of monoclonal antibodies reactive with the** α**,** β**, and** γ_c **IL-2R chains.** The cells were incubated at 37°C with 200 pM ^{125}I-IL-2 in the absence of antibody (●), or in the presence of anti-IL-2Rα-7G7B6 ($\triangle$), or anti-IL-2Rβ-341 (▼), or anti-IL-2Rγ TUG4h (■). (Redrawn from: Hemar, A. *et al.* 1995. *J. Cell Biol.* **129**:55–64.)

radiolabeled IL-2 from the cell surface in the presence of non-IL-2 competitive MoAbs reactive with the three distinct IL-2R chains are shown.[5] IL-2 is internalized very rapidly, with a $t_{1/2}$ equal to 15 minutes. Since IL-2 dissociates from high affinity IL-2 binding sites with a $t_{1/2}$ equal to 45 minutes, the internalization rate essentially determines the duration that IL-2Rs can signal. The degradation of radiolabeled IL-2 is shown in Fig. 8.4B, and reveals that ~50% is degraded within 60 minutes after binding. Thus, the kinetics of binding, internalization and degradation of IL-2 are important for determining the ultimate tempo, magnitude and duration of the IL-2-dependent clonal expansion of antigen-selected T cells.

Dautry-Varsat's group showed that like many peptide/protein hormone-receptor systems that have been studied, IL-2 binding to high affinity IL-2Rs promotes a ligand-dependent acceleration of the internalization of the IL-2Rs.[4] Thus, within two hours of IL-2 exposure a new steady state of IL-2R density is achieved, which is ~50% of the initial IL-2R density. Since IL-2R density is one of the parameters that determines the rate of the proliferative response, this phenomenon of accelerated IL-R internalization is important. Dautry-Varsat's group also showed that the three chains of the IL-2R are internalized together with IL-2 as a quaternary complex, but then the receptor chains are sorted separately within the cell.[5] The IL-2Rα chain is internalized only when part of high affinity receptors that are only formed when IL-2 is present. Once internalized, IL-2Rα chains are recycled to the cell surface, thereby maintaining the excess of IL-2Rα chains compared with the β and γ_c chains. By comparison, the β and γ_c chains traffic to lysosomes and are degraded, thereby providing the mechanism for the ligand-dependent accelerated disappearance in the surface density of the high affinity IL-2Rs.

In further exploration of IL-2R endocytosis, Dautry-Varsat's group discovered that high affinity IL-2Rs are internalized via a novel clathrin-independent pathway that had previously only been found to internalize toxins and some viruses.[6,7] Therefore, IL-2R endocytosis was the first physiological ligand-receptor pair to be found not to proceed via clathrin-coated pits. More recent work characterizing the molecules involved in this pathway has indicated that it requires both RhoA and Rac1, which actually antagonize clathrin-dependent endocytosis.[8] Also, the serine/threonine kinases Pak1 and Pak2 are stimulated by Rac1, and cortactin is a downstream target of the Paks. Thus, there is differential regulation of the two endocytic pathways, which share important factors such as dynamin, actin and cortactin, with Rac1-Pak1-Pak2 acting as upstream regulators that switch on the clatherin-independent pathway. Obviously, this pathway is probably involved in the endocytosis of all γ_c cytokines. Accordingly, there will be more investigations forthcoming as to how this internalization pathway functions, and whether there are any differences between the internalization of the receptors for each γ_c family member, as well as whether other members of the interleukin receptor family also are internalized via this mechanism.

References

1. Robb, R.J., Munck, A., and Smith, K.A. (1981) T cell growth factor receptors: quantitation, specificity, and biological relevance. *J. Exp. Med.* **154**:1455–1474.
2. Smith, K.A., and Cantrell, D.A. (1985) Interleukin 2 regulates its own receptors. *Proc. Natl. Acad. Sci. USA* **82**:864–868.
3. Duprez, V., and Dautry-Varsat, A. (1986) Receptor-mediated endocytosis of interleukin-2 in a human tumor cell line. *Proc. Natl. Acad. Sci. USA* **261**:15450–15454.
4. Duprez, V., Cornet, V., and Dautry-Varsat, A. (1988) Down-regulation of high affinity interleukin-2 receptors in a human tumor T cell line. *J. Biol. Chem.* **263**:12860–12865.
5. Hemar, A., Subtil, A., Lieb, M., Morelon, E., Hellio, R., and Dautry-Varsat, A. (1995) Endocytosis of interleukin-2 receptors in human T lymphocytes: distinct intracellular localization and fate of the receptor alpha, beta, and gamma chains. *J. Cell Biol.* **129**:55–64.
6. Lamaze, C., Dujeancourt, A., Baba, T., Lo, C., Benmerah, A., and Dautry-Varsat, A. (2001) Interleukin-2 receptors and detergent-resistant membrane domains define a clathrin-independent endocytic pathway. *Mol. Cell.* **7**:661–671.
7. Sauvonnet, N., Dujeancourt, A., and Dautry-Varsat, A. (2005) Cortactin and dynamin are required for the clathrin-independent endocytosis of gamma-c cytokine receptor. *J. Cell. Biol.* **168**:155–163.
8. Grassart, A., Dujeancourt, A., Lazarow, P., Dautry-Varsat, A., and Sauvonnet, N. (2008) Clathrin-independent endocytosis used by the IL-2 receptor is regulated by Rac1, Pak1 and Pak2. *EMBO reports* **9**:356–362.

Chapter 9

The Molecular Basis for Quantal IL-2/IL-2R Signaling of Cell Cycle Progression — IL-2 Receptor Signaling via the Jak/Stat Pathway

When each of the three IL-2R chains were identified,[1-4] and their cDNAs cloned and sequenced,[5-8] it was surprising that the sequences of the cytoplasmic domains did not reveal any known enzymatic motifs capable of generating biochemical reactions in the cell interior. Even before the identification of the IL-2R chains, experiments had revealed very rapid IL-2/IL-2R-dependent phosphorylation of multiple cytoplasmic proteins.[9] In addition, it had already been determined that the cytoplasmic domains of the epidermal growth factor receptor (EGFR) and the platelet-derived growth factor receptor (PDGFR) contained sequences consistent with PTK activity. These are now called receptor tyrosine kinases (RTK).

Fortunately, investigators searching for new and novel kinases by homology cloning identified a new family of cytoplasmic PTKs that were termed Janus kinases (JAKs), because they contained two kinase domains organized in opposite orientation. Thus, the mythical Roman god Janus who has two faces, one looking forward and the other backward, was used to name this family, which ultimately was found to have four members: JAKs 1, 2 and 3, and Tyk-2.[10-12] The JAKs were originally found to be phosphorylated rapidly upon interferon (IFN) signaling,[13] and soon thereafter were shown to be involved in signaling via several of the hematopoietic cytokines, including the erythrocyte growth factor, erythropoietin (EPO),[14] the mast cell growth factor IL-3,[15] and granulocyte-colony stimulating factor (G-CSF).[16]

Even before the JAKs were identified, experiments by James Darnell's group uncovered a new family of cytoplasmic molecules that they termed signal transducers and activators of transcription (STAT), because these molecules were shown to convey IFN signals from the membrane to the nucleus.[17-22] Seven members of this molecular family were rapidly identified, and ingenious somatic cell genetic approaches by George Stark and Ian Kerr with their co-workers showed that the JAKs phosphorylate and activate the STATs[13,23] (review in Ref. 24).

STAT5 was first identified to be involved in prolactin signaling in sheep mammary tissue,[25] and two isoforms were subsequently found to be 95% identical,[26] so that they are termed STAT5a and STAT5b. Experiments by Carol Beadling working in Doreen Cantrell's group showed that the IL-2R activates JAK1 and JAK3, which then phosphorylate and activate STAT5.[27] It is especially noteworthy that by comparison, they found that the TCR does not activate either the JAK family kinases or any of the STAT family molecules. However, TCR stimulation via mitogens or anti-CD3 activates the expression of JAK3 protein.[28] Subsequent experiments revealed that the IL-2R β chain associates with JAK1, while the IL-2R γ_c chain associates with JAK3.[29,30] Upon IL-2 binding, which comes to a steady state within 10–15 minutes when used at IL-2R saturating concentrations (i.e. > 100 pM), IL-2 stimulates rapid phosphorylation of both JAK1 and JAK3.

The lack of structural information for the cytoplasmic domains of the IL-2R chains prevents any insight on how the IL-2R β and γ_c chains interact with the JAK1 and JAK3 molecules, as well as to how ligand binding to the external domains of the IL-2R activates the kinase activities. However, it has become accepted, based on co-precipitation experiments, that even in the absence of ligand the cytoplasmic domains of the IL-2R β and γ_c chains are already associated in non-covalent complexes with the JAKs. Thus, bringing the cytoplasmic domains together by IL-2 binding to the IL-2R external domains somehow leads to conformational changes in IL-2R cytoplasmic domains, which activate the JAKs by releasing their C-terminal catalytic domains from suppression by their pseudokinase domains, thereby initiating enzymatic activity.

The complete three-dimensional structure of a JAK molecule has yet to be solved. However, based upon the primary structure information, the JAK molecules are known to have four domains, including an N-terminal FERM homology domain, which has been implicated in binding to the receptor chain, followed by an SH-2 domain, the pseudokinase domain and the C-terminal catalytic domain.[31] All of the data accumulated thus far are consistent with the hypothesis that the formation of a very stable macromolecular quaternary IL-2/IL-2R complex occurs upon IL-2 binding to the extracellular chains, which continues to signal as long as IL-2 remains bound. Accordingly, as already mentioned, the $t_{1/2}$ of 15 minutes for the measured internalization rate of radiolabeled IL-2 is a good estimate of the median functional lifetime of a ligand bound signaling IL-2/IL-2R/JAK macromolecular complex. Presumably, only upon internalization will the acid pH of the receptosome promote dissociation of IL-2 from the external domains of the IL-2R chains, disrupting and dissolving the macromolecular complex, and thereby extinguishing signaling.

The JAK-mediated PTK activity results in phosphorylation of tyrosine residues on the JAKs themselves, as well as tyrosine residues on the IL-2R β chain, which then serve as docking sites for intracellular messenger molecules. Mutational analysis has shown that the IL-2R β chain Y510 serves as a docking residue for STAT5, and Y338 serves as a docking residue for Shc.[32] Because IL-2-promoted T cell cycle progression depends upon achieving a critical number of IL-2/IL-2R interactions, it follows that this digital information must somehow be transferred to the interior of the cell. Thus, the number of IL-2/IL-2R interactions must be transferred into a critical number of intracellular messenger molecules. Since the IL-2/IL-2R activates JAK1 and JAK3, it is logical to hypothesize that the number of these molecules activated per unit time ultimately determines cell cycle progression. However, these parameters have yet to be determined experimentally.

Based upon experiments performed during the past 15 years, what happens on the inside of the cell after IL-2/IL-2R interactions on the cell surface is now known. Three signaling pathways are activated, all via the JAK1/3 PTK phosphorylation of the IL-2R β chain: STAT5, PI3K, and MAPK. Because the STAT molecules traffic from the membrane directly to the transcription machinery of the nucleus, the STAT molecules have received a great deal of attention since their discovery. Structure activity relationship (SAR) studies of the STAT family molecules have revealed a great deal about how these molecules function. All seven mammalian STAT proteins share six structural regions: an N-terminal domain (ND), a coiled-coil domain (CC), a DNA binding domain (DBD), a linker domain (L), an Src homology 2 (SH2) domain, and a transcriptional activation domain (TAD) at the carboxy terminus, as shown schematically by Mertens and Darnell.[33] The molecule becomes activated when a single tyrosine residue within the activation domain at the C-terminus (Y ~ 700) becomes phosphorylated by the JAKs. Prior to activation, unphosphorylated STAT proteins form stable dimers in the cytoplasm, which are structurally distinct from the tyrosine-phosphorylated STAT dimers that form after activation. Unphosphorylated STAT5 molecules in the cytoplasm dimerize in an anti-parallel fashion through a CC:DBD interaction,[34] which leaves the SH-2 domains and C-terminal activation domains at opposite ends of the dimer.

JAK-mediated phosphorylation of residue Y510 on the cytoplasmic domains of the IL-2R β chain serves as docking sites for the SH-2 domains of the STAT5 molecules and a reorientation of the STAT dimers, placing them in a parallel fashion.[33] This reorientation allows the JAKs to phosphorylate the critical C-terminal residue ~Y700, which then solidifies the parallel STAT dimer by reciprocally binding to the SH-2 domains, and thereby promoting the dissociation of the parallel dimer from the docking sites on the receptor chains. This released parallel STAT dimer then enters the nucleus via importin-α-dependent transport and binds to STAT5-specific DNA target response elements (RE).

It is important to realize that although we now know that this is *what* happens, we have no data to indicate exactly *how* it happens.

Thus we do not know the relative concentrations of the relevant molecules involved in these intermolecular associations and dissociations, nor do we know the equilibrium and kinetic binding constants, or the energetics of the reactions, in the same way that we have determined these parameters for IL-2 interaction with the IL-2R chains, and the interactions between the IL-2R chains, with or without bound IL-2. Also, no one has examined these parameters at the level of individual cells, and examined whether there is variability in the concentrations of crucial molecules among cells within a population, and whether they function in an analogue or digital fashion. However, at this point it must be assumed that as long as IL-2 remains bound to the external domains of the IL-2R chains, the cytoplasmic chains will continue to form stable macromolecular cytoplasmic complexes, and continue to generate activated phosphorylated parallel STAT5 dimers (pSTAT5$_2$), which continue to enter the nucleus and bind DNA response elements (RE).[33] Accordingly, the critical number of IL-2/IL-2R complexes formed per unit time is the initial stable macromolecular complex that creates the rate-limiting intracellular macromolecular complexes that initiate the digital cellular response, and move the cell beyond the R-point to quantal cellular DNA synthesis and mitosis.

Once the activated pSTAT5$_2$ molecules enter the nucleus, in order to drive gene expression the STAT proteins must cooperate with numerous other transcriptional activating factors (e.g. IRFs, Sp1, Jun, Fos, NF-κB, glucocorticoid R), and co-activators with functions in chromatin remodeling (e.g. p300/CBP, PCAF, GCN5, BRG1, HDACs) or in the formation of pre-initiation complexes.[33] Moreover, maximal transcriptional activation requires serine phosphorylation of a conserved motif in the STAT C-terminal activation domain. Should activated pSTAT5$_2$ dissociate from DNA, nuclear protein phosphatases (N-PTP) remove the N-terminal phosphate, favoring the dissociation of the parallel homodimer and the reformation of the anti-parallel homodimer, which is then free for transport back to the cytoplasm where it can be reactivated. Accordingly, the STAT molecules are stable throughout the activation/deactivation cycle and recycle from the cytoplasm to the nucleus depending on the ligand/receptor reaction at the cell surface. Thus, the number of STAT molecules within each cell, as well as the relative numbers of unactivated versus activated molecules, becomes of utmost importance in terms of perpetuating the signal. These parameters are the most critical in determining which genes, and the duration that STAT-responsive genes will be transcribed. Thus, the ligand-dependent activation of the STAT molecules, and the rate of decay of $pSTAT5_2$ upon removal of IL-2R-bound IL-2 ultimately determine the number of STAT5-directed gene products generated. In this way, the digital read-out from the cell surface generated IL-2/IL-2R-derived signals is preserved.

Even before the IL-2 activation of STAT5 was discovered, IL-2 was shown to be responsible for promoting a marked increase in cell size (i.e. lymphocyte blastogenesis) via activating RNA and protein synthesis prior to progression through the R-point to the S-phase.[35] In addition, IL-2 was found to activate the expression of specific genes, independently from those activated via the cell cycle "competence" signals generated by the TCR.[36] Subsequently, Julia Turner documented IL-2-dependent blastogenic transformation using flow cytometry.[37] Moreover, she found that the expression of cyclin D2 and cyclin D3 mRNA precedes the G_1-S-phase transition in an IL-2 concentration and time-dependent manner. In a G_0/G_1 synchronized population of human IL-2R$^+$ T cells, IL-2 promoted a sequential expression of cyclin D2, first detectable after three hours, with a peak expression after six hours and a gradual decline to low levels by the onset of S-phase at 21.5 hours. Cyclin D3 expression followed cyclin D2 expression, first becoming detectable after six hours, with peak expression at 21.5 hours coinciding with S-phase transition, remaining elevated throughout S-phase to 32 hours, then falling to basal levels as the cells underwent cytokinesis and returned to G_1. IL-2 also induced cyclin E expression, which also first became detectable after six hours, peaking between 11 and 24 hours coinciding with G_1-S-phase transition, remaining elevated during S-phase and declining thereafter.

In these experiments, it is noteworthy that phorbol dibutyrate (PdBu), which was used to mimic TCR activation to promote IL-2R

α chain expression, did not increase the basal levels of the G_1 cyclins, or promote cell cycle progression without the addition of IL-2. Accordingly, these results are entirely consistent with the interpretation that the TCR confers cell cycle "competence" on quiescent T cells, while IL-2/IL-2R signals cell cycle "progression". This interpretation was further supported by experiments in which the cyclosporine-A (CsA)-mediated block of T cell proliferation, which is known to occur via inhibition of IL-2 and IL-2R gene expression, could be shown to have no effect on G_1 cyclin expression, provided IL-2 was supplied exogenously. Thus, CsA blocks TCR-promoted cell cycle "competence" but not IL-2-mediated cell cycle "progression." By comparison, another inhibitor of T cell proliferation, rapamycin, also did not block IL-2-dependent G_1 cyclin expression, but did block IL-2-dependent cell cycle progression, thereby indicating that there are at least two distinct pathways signaled by the IL-2/IL-2R interaction that promote G_1 progression. As detailed in the next chapter, we now know that rapamycin blocks the IL-2-promoted decrease in the p27^{Kip} cell cycle inhibitor, instead of regulating the expression of the cyclins.[38]

It is also noteworthy that these experiments were performed using whole T cell populations, so that the analysis of cyclin expression was necessarily only semi-quantitative, precluding any correlation between individual cell IL-2R numbers triggered and the number of G_1 cyclin molecules expressed. However, even at this level of experimentation, the results are consistent with a direct connection between the number of IL-2Rs triggered at the cell surface and the magnitude and duration of G_1 cyclin gene expression. Even so, to proceed beyond these cell population experiments, it will be necessary to quantify individual cell levels of the cyclin proteins. Fortunately, this has been done, particularly by Zbigniew Darzynkiewicz and co-workers, who have documented cyclin expression and cell cycle position with quantitative DNA determination using flow cytometry.[39] Thus, cyclin D1 expression is confined to G_1 in proliferating cells, while cyclin E expression occurs at the G_1-S-phase transition and throughout S-phase. Cyclin A expression is minimal during G_1, begins during S-phase and increases progressively as the cells advance toward and enter G_2. As cells enter pro-metaphase, cyclin A is rapidly degraded. By comparison, expression of cyclin B1 is limited to late S-phase and to cells with a G_2/M content. Given the capacity to monitor both the number of IL-2Rs at the cell surface, the number of activated $pSTAT5_2$ molecules in the nucleus, and the number of G_1 cyclins expressed, it is now possible to begin to determine how the system works.

When G_1 synchronized T cells are stimulated by IL-2 to enter the cell cycle, transcription of cyclin D2 followed by cyclin D3 and then cyclin E occurs, apparently in a digital fashion according to the number of triggered IL-2Rs. The D and E cyclins then assemble with their catalytic partners, CDK4/6 and CDK2 respectively, as the cells progress through G_1. Assembled cyclin D-CDK complexes then enter the cell nucleus where they must be phosphorylated by a CDK-activating kinase (CAK) to be able to phosphorylate their target protein substrates, in particular the Retinoblastoma (Rb) proteins, which function to suppress the E2F transcription factors during interphase. As these transcription factors are released from Rb inhibition, they coordinately activate cyclin E and the genes necessary for nucleotide synthesis and DNA replication (for review, see Ref. 40). In addition, cyclin E/CDK2 complexes activate the origins of DNA replication, thus, initiating DNA synthesis. These cyclin D/CDK4/6 and cyclin E/CDK2 effects are thought to be responsible for passing the R-point, and for the G_1-S-phase transition. Accordingly, D-type cyclins act as IL-2/IL-2R sensors with cyclin D transcription, assembly, nuclear transport, and turnover each being IL-2R-dependent quantal steps. Because the cyclin D2 gene has been shown to have STAT5 REs and to be regulated by pSTAT5, this IL-2R/STAT5/cyclin D2 pathway represents the most straightforward receptor to gene pathway to begin to understand the biochemistry and biophysics of how a cell "counts."[41,42] In this regard, it perhaps goes without saying that should this counting system be circumvented by a mutational usurpation of the digital process, an autonomously proliferating cell could result (a.k.a. cancer).

It is also noteworthy that gene deletion studies have shown that all three D cyclins[43] as well as both CDK4 and CDK6[44] can be deleted and embryogenesis still progresses to midgestation. In addition,

MEFs explanted from day E13.5 can be grown *in vitro*, but progress through the cell cycle more slowly than WT MEFs. Moreover, cell cycle progression by these gene-deleted cells requires greater concentrations of serum (as a source of growth factors) than WT cells. The embryos die *in utero* due to the lack of formation of blood cells, so that these data are consistent with a requirement of the G_1 cyclins for a rapid response to mitogenic hematopoietic cytokines. Even so, it appears that G_1 progression can occur in the absence of the D cyclin/CDK activity, presumably mediated by cyclin E/CDK2 activity. However, the fact that the gene deleted MEFs are still dependent on the growth factors in serum for their proliferative responses indicates that mitogenic factors working at the cell surface control cell cycle progression of all cells, and that different cell types may utilize different cytokine/receptor systems and different cyclin/CDK complexes to signal cell cycle progression.

References

1. Leonard, W.J., Depper, J.M., Uchiyama, T., Smith, K.A., Waldmann, T.A., and Greene, W.C. (1982) A monoclonal antibody that appears to recognize the receptor for human T-cell growth factor; partial characterization of the receptor. *Nature* **300**:267–269.
2. Sharon, M., Klausner, R.D., Cullen, B.R., Chizzonite, R., and Leonard, W.J. (1986) Novel interleukin-2 receptor subunit detected by cross-linking under high affinity conditions. *Science* **234**:859–863.
3. Teshigawara, K., Wang, H.M., Kato, K., and Smith, K.A. (1987) Interleukin 2 high-affinity receptor expression requires two distinct binding proteins. *J. Exp. Med.* **165**:223–238.
4. Takeshita, T., Ohtani, K., Asao, H., Kumaki, S., Nakamura, M., and Sugamura, K. (1992) An associated molecule, p64, with IL-2 receptor beta chain. Its possible involvement in the formation of the functional intermediate-affinity IL-2 receptor complex. *J. Immunol.* **148**:2154–2158.
5. Leonard, W.J., Depper, J.M., Crabtree, G.R., Rudikoff, S., Pumphrey, J., Robb, R.J., Kronke, M., Svetlik, P.B., Peffer, N.J., Waldmann, T.A., *et al.* (1984) Molecular cloning and expression of cDNAs for the human interleukin-2 receptor. *Nature* **311**:626–631.
6. Nikaido, T., Shimizu, A., Ishida, N., Sabe, H., Teshigawara, K., Maeda, M., Uchiyama, T., Yodoi, J., and Honjo, T. (1984) Molecular cloning of cDNA encoding human interleukin-2 receptor. *Nature* **311**:631–635.

7. Hatakeyama, M., Tsudo, M., Minamoto, S., Kono, T., Doi, T., Miyata, T., Miyasaka, M., and Taniguchi, T. (1989) Interleukin-2 receptor beta chain gene: generation of three receptor forms by cloned human alpha and beta chain cDNA's. *Science* **244**:551–556.
8. Takeshita, T., Asao, H., Ohtani, K., Ishii, N., Kumaki, S., Tanaka, N., Munakata, H., Nakamura, M., and Sugamura, K. (1992) Cloning of the gamma chain of the human IL-2 receptor. *Science* **257**:379–382.
9. Merida, I., and Gaulton, G. (1990) Protein tyrosine phosphorylation associated with activation of the interleukin-2 receptor. *J. Biol. Chem.* **265**:5690–5694.
10. Wilks, A. (1989) Two putative protein-tyrosine kinases identified by application of the polymerase chain reaction. *PNAS* **86**:1603–1607.
11. Firmbach-Kraft, I., Byers, M., Shows, T., Dalla-Favera, R., and Krolewski, J. (1990) *tyk2*, prototype of a novel class of non-receptor tyrosine kinase genes. *Oncogene* **5**:1329–1336.
12. Wilks, A., Harpur, A., Kurban, R., Ralph, S., Zurcher, G., and Ziemiecki, A. (1991) Two novel protein-tyrosine kinases, each with a second phosphotrans-ferase-related catalytic domain, define a new class of protein kinase. *Mol. Cell. Biol.* **11**:2057–2065.
13. Velazquez, L., Fellous, M., Stark, G., and Pellegrini, S. (1992) A protein tyrosine kinase in the interferon alpha/beta signaling pathway. *Cell* **70**:313–322.
14. Witthuhn, B., Quelle, F., Silvennoinen, O., Yi, T., Tang, B., Miura, O., and Ihle, J. (1993) JAK2 associates with the erythropoietin receptor and is tyrosine phosphorylated and activated following stimulation with erythropoietin. *Cell* **74**: 227–236.
15. Silvennoinen, O., Witthuhn, B., Quelle, F., Cleveland, J., Yi, T., and Ihle, J. (1993) Structure of the murine Jak2 protein-tyrosine kinase and its role in interleukin 3 signal transduction. *PNAS* **90**:8429–8433.
16. Shimoda, K., Iwasaki, H., Okamura, S., Ohno, Y., Kubata, A., Arima, F., Otsuka, T., and Niho, Y. (1994) G-CSF induces tyrosine phosphorylation of the JAK2 protein in the human myeloid G-CSF responsive and proliferative cells, but not in mature neutrophils. *Biochem. Biophys. Res. Commun.* **203**:922–928.
17. Larner, A., Jonak, G., Cheng, Y.-S., Korant, B., Knight, E., and Darnell, J.E., Jr. (1984) Transcriptional induction of two genes in human cells by beta interferon. *PNAS* **81**:6733–6737.
18. Larner, A., Chaudhuri, A., and Darnell, J.E., Jr. (1986) Transcriptional induction by interferon. *J. Biol. Chem.* **261**:453–459.
19. Fu, X., Kessler, D., Veals, S., Levy, D., and Darnell, J. (1990) ISGF3, the transcriptional activator induced by interferon alpha, consists of multiple interacting polypeptide chains. *PNAS* **87**:8555–8559.
20. Fu, X., Schindler, C., Improta, T., Aebersold, R., and Darnell, J.E., Jr. (1992) The proteins of ISGF-3, the interferon {alpha}-induced transcriptional activator, define a gene family involved in signal transduction. *PNAS* **89**:7840–7843.

21. Sadowski, H.B., Shuai, K., Darnell, J.E., Jr., and Gilman, M.Z. (1993) A common nuclear signal transduction pathway activated by growth factor and cytokine receptors. *Science* 261:1739–1744.

22. Schindler, C., Shuai, K., Prezioso, V., and Darnell, J. (1992) Interferon-dependent tyrosine phosphorylation of a latent cytoplasmic transcription factor. *Science* **257**:809–813.

23. Pelligrini, S., John, J., Shearer, M., Kerr, I., and Stark, G. (1989) Use of a selectable marker regulated by alpha interferon to obtain mutations in the signaling pathway. *Mol. Cell. Biol.* 9:4605–4612.

24. Darnell, J., Kerr, I., and Stark, G. (1994) Jak-STAT pathways and transcriptional activation in response to IFNs and other extracellular signaling proteins. *Science* **264**: 1415–1421.

25. Wakao, H., Gouilleux, F., and Groner, B. (1994) Mammary gland factor (MGF) is a novel member of the cytokine regulated transcription factor gene family and confers prolactin response. *EMBO J.* **13**:2182–2191.

26. Liu, X., Robinson, G.W., Gouilleux, F., Groner, B., and Hennighausen, L. (1995) Cloning and expression of Stat5 and an additional homologue (Stat5b) involved in prolactin signal transduction in mouse mammary tissue. *PNAS* **92**:8831–8835.

27. Beadling, C., Guschin, D., Witthuhn, B., Ziemiecki, A., Ihle, J., Kerr, I., and Cantrell, D. (1994) Activation of JAK kinases and STAT proteins by interleukin-2 and interferon alpha, but not the T cell antigen receptor, in human T lymphocytes. *EMBO J.* **13**:5605–5615.

28. Johnston, J., Kawamura, M., Kirken, R., Chen, Y.-Q., Blake, T., Shibuya, K., Ortaldo, J., McVicar, D., and O'Shea, J. (1994) Phosphorylation and activation of the JAK-3 Janus kinase in response to interleukin-2. *Nature* 370:151–153.

29. Russell, S., Johnston, J., Noguchi, M., Kawamura, M., Witthuhn, B., Silvennoinen, O., Goldman, A., Schmalsteig, F., Ihle, J., O'Shea, J., et al. (1994) Interaction of IL2R beta and gamma-c chains with JAK1 and JAK3: implications for XSCID and XCID. *Science* 266:1042–1045.

30. Miyazaki, T., Kawahara, A., Fujii, H., Nakagawa, Y., Minami, Y., Liu, Z.J., Oishi, I., Silvennoinen, O., Witthuhn, B.A., Ihle, J.N., et al. (1994) Functional activation of Jak1 and Jak3 by selective association with IL-2 receptor subunits. *Science* 266:1045–1047.

31. Pesu, M., Laurence, A., Kishore, N., Zwillich, S., Chan, G., and O'Shea, J. (2008) Therapeutic targeting of Janus kinases. *Immunol. Rev.* **223**:132–142.

32. Friedman, M., Migone, T., Russell, S., and Leonard, W. (1996) Different IL-2 receptor beta chain residues couple to at least two signaling pathways and synergistically mediate interleukin-2-induced proliferation. *Proc. Natl. Acad. Sci. USA* **93**:2077–2082.

33. Mertens, C., and Darnell, J.J. (2007) Snapshot: JAK-STAT signaling. *Cell* **131**:612.

34. Neculai, D., Neculai, A., Verrier, S., Straub, K., Klumpp, K., Pfitzer, E., and Becker, S. (2005) Structure of unphosphorylated STAT5a dimer. *J. Biol. Chem.* **280**:40782–40787.
35. Stern, J.B., and Smith, K.A. (1986) Interleukin-2 induction of T-cell G1 progression and c-myb expression. *Science* **233**:203–206.
36. Beadling, C., Johnson, K.W., and Smith, K.A. (1993) Isolation of interleukin 2-induced immediate-early genes. *Proc. Natl. Acad. Sci. USA* **90**:2719–2723.
37. Turner, J. (1993) IL2-dependent induction of G1 cyclins in primary T cells is not blocked by rapamycin or cyclosporin A. *Int. Immunol.* **10**:1199–1209.
38. Nourse, J., Firpo, E., Flanagan, W., Coats, S., Polyak, K., Lee, M., Massague, J., Crabtree, G., and Roberts, J. (1994) Interleukin-2-mediated elimination of the p27Kip1 cyclin-dependent inhibitor prevented by rapamycin. *Nature* **372**:570–573.
39. Darzynkiewicz, Z., Juan, G., and Bedner, E. (1999) Determining cell cycle stages by flow cytometry. *Current Protocols in Cell Biol.* **8(1)**, **Unit 4**:1–18.
40. Nevins, J., Leone, G., DeGregori, J., and Jakoi, L. (1997) Role of the Rb/E2F pathway in cell growth control. *J. Cell. Physiol.* **173**:233–236.
41. Smith, K.A. (1995) Cell growth signal transduction is quantal. *Ann. New York Acad. Sci.* **766**:263–271.
42. Smith, K. (1997) Why do cells count? In *Nonlinear Cooperative Phenomena in Biological Systems*. L. Matsson, editor. World Scientific Publishing Co. Pte. Ltd. Singapore, 13–19 pp.
43. Kozar, K., Ciemerych, M., Rebel, V., Shigematsu, H., Zagozdon, A., Sicinska, E., Qunyan, Y., Yu, Q., Bhattacharya, S., Bronson, R., *et al.* (2004) Mouse development and cell proliferation in the absence of D cyclins. *Cell* **118**:477–491.
44. Malumbres, M., Sotillo, R., Santamaria, D., Galan, J., Cerezo, A., Ortega, S., Dubus, P., and Barbacid, M. (2004) Mammalian cells cycle without the D-type cyclin-dependent kinases cdk4 and cdk6. *Cell* **118**:493–504.

Chapter 10

The Molecular Basis for Quantal IL-2/IL-2R Signaling of Cell Cycle Progression — IL-2 Receptor Signaling via Phosphorylation of IL-2R β Chain Y338

In addition to Y510, JAK1 and JAK3 also phosphorylate Y338 on the IL-2R β chain, which serves as a docking site for the adapter protein, Shc.[1,2] Shc is ubiquitously expressed and exists in three isoforms of 66, 52, and 48 kDa.[3,4] The Shc protein is composed of three domains, an N-terminal domain that interacts with proteins containing pYs, a (gly/pro)-rich collagen homology domain, and an Src-homology-2 (SH2) domain, which binds to Y338 on the IL-2R β chain. Once recruited to the IL-2R β chain, Shc activates at least two downstream pathways, the PI3K (phosphoinositide 3-kinase) pathway, and the MAPK (mitogen activated protein kinase) pathway.[4] Shc activates the PI3K pathway by recruiting the adapter protein Grb2, which in turn recruits the adapter protein Gab2, followed by the p85 PI3K regulatory subunit. Formation of the Shc/Grb2/Gab2/p85 complex ultimately leads to activation of the catalytic subunit of PI3K, p110, which converts phosphatidylinositol 4,5-bisphosphate into the lipid second messenger phosphatidylinositol 3,4,5-triphosphate (PIP_3) in the cell membrane. PIP_3 then recruits several mediators to the cell membrane, which are proteins containing pleckstrin homology (PH) domains, such as 3-phosphoinositide-dependent kinase-1 (PDK1), PKB (a.k.a. Akt), and the 70-kDa ribosomal S6 kinases (S6Ks). PKB regulates glucose uptake and protein synthesis by increasing the kinase activity of the mammalian Target of Rapamycin (mTOR).

Detailed studies using a model system of IL-3 and IL-3-dependent cells have shown that activated PKB prevents IL-3-dependent withdrawal apoptosis via the maintenance of surface transporters for several critical metabolites, including glucose, amino acids, low-density lipoprotein, and iron.[5] Activated PKB maintains these transporters on the cell surface in the absence of IL-3 through a mTOR-dependent mechanism; in that the mTOR inhibitor rapamycin diminishes PKB-dependent increases in cell size, mitochondrial membrane potential, and cell survival. Thus, cytokines like IL-2 and IL-3 control cellular size and survival by regulating cellular access to extracellular nutrients, in part by activating PI3K, which controls the activity of PKB and mTOR. Accordingly, the mechanisms controlling "lymphocyte blastogenesis" (i.e. increase in cell size and immature morphology), originally observed and documented by Nowell, are now finally known at the molecular level, at least in outline.

It appears that continuous IL-2 triggering of the IL-2R is necessary to sustain the large increase in cell size that occurs upon T cell activation, in that withdrawal of IL-2 results in a rapid decrease in cell size followed by apoptosis. Cantrell and co-workers have shown that antigen-primed T cells respond to IL-2 by a marked increase in cell size, as detected by phase contrast microscopy and by flow cytometric analysis of forward and side scatter.[6] By comparison, IL-15 does not lead to a similar size increase, even though IL-15 also uses both the IL-2R β and γ chains to signal. A comparison of PI3K signaling by IL-2- vs. IL-15-stimulated CD8[+] T cells revealed that IL-15 activation of the PI3K pathway is transient and weaker than IL-2 stimulation. Accordingly, detailed quantitative studies are now required to relate the numbers of ligand/receptor triggers at the cell surface to the number, concentration and kinds of intracellular biochemical molecules and pathways activated with the consequent numbers of transport molecules placed on the cell surface. It is noteworthy that the differences between the metabolic changes promoted by IL-2 and by IL-15 reflects what is known about their physiologic roles in the immune system. Thus, IL-2 promotes a rapid proliferative expansion of antigen-selected T cells, while IL-15 is important for NK cell development and the maintenance of NK cells and the memory

CD8[+] T cell population in the periphery by providing a slow proliferative stimulus.[6]

The metabolic changes promoted by IL-2 are important and informative when viewed from the context of malignancy. It has recently become appreciated that the long known propensity of cancer cells to utilize glycolysis for energy production, even in the presence of oxygen (known as the Warburg effect, see Ref. 7), may be due to the persistent activation of the PI3K/PKB/mTOR pathway.[8,9] Thus, most tumor cells have a substantial reserve capacity to produce ATP by oxidative phosphorylation that can be demonstrated by suppressing glycolysis. However, because malignancy results from autonomous activation of the signaling pathways normally under the control of cytokine/receptor systems, persistent activation of the PI3K/PKB/mTOR pathway favors the rapid production of ATP via glycolysis.

In addition to nutrient transporters, PI3K activation also regulates the surface expression of lymphocyte adhesion and chemokine receptors, thereby markedly altering the migratory capacity of T cells. Cantrell's group has shown that the key lymph node-homing receptors CD62L (L-selectin) and CCR7, which are highly expressed on naive and memory T cells and promote homing to peripheral lymph nodes, are down-regulated after TCR and cytokine activation, thereby favoring the migration of antigen/cytokine-primed T cells to sites of inflammation.[10] Of interest, CD62L down-regulation occurs through ectodomain proteolysis mediated via the p110δ catalytic subunit of PI3K, which acts through MAPKs. In addition, p110δ suppresses CD62L transcription via an mTOR-dependent inhibition of expression of the transcription factor KLF2, which activates CD62L transcription. Moreover, KLF2 also regulates the transcription of the CCR7 gene, so that PI3K activation of mTOR also results in suppression of CCR7 expression, a pre-requisite for migration from the secondary lymphoid organs to peripheral tissues. Accordingly, the PI3K signaling pathway controls both T cell metabolism and cell size, as well as T cell migration by regulating the expression of cell surface nutrient transporters and homing receptors.

It is noteworthy that the SARs of the interactions between Y338 from the IL-2R β chain and the second messengers beginning with

Shc that are important for activating both the PI3K and the MAPK pathways have yet to be determined completely. Moreover, the kinds of biophysical studies performed with IL-2 binding to the IL-2R chains, such as SPR and ITC, which are important for understanding how the biochemical signals are generated and propagated, have not been performed. Thus, very important studies involving single cell analyses, and quantification of the various signaling molecules involved in the reactions must be performed to understand how the phenotypic changes in the cells are generated. This is especially true in regards to the mTOR kinase molecules, an area that is still rapidly evolving.

With regard to the activation of cell cycle progression, soon after the identification of the D cyclins and the cyclin-dependent kinase inhibitor p27^{Kip1}, experiments revealed that IL-2-promoted T cell cycle progression is accompanied by a decrease in p27^{Kip1} expression, thereby facilitating G_1 progression.[11,12] Moreover, this decrease in p27^{Kip1} expression is inhibited by rapamycin, placing the mechanism of this IL-2 effect within the PI3K/PKB/mTOR pathway. These data indicate that rapamycin does not block G_1 progression by inhibiting the expression of the D cyclins, which was first documented by Julia Turner,[13] but rather by preventing the disappearance of the CKI.

Initially the IL-2 signaling effect on p27^{Kip1} expression was suggested to be mediated by an enhanced rate of p27^{Kip1} degradation. However, subsequent experiments have now shown that cytokine activation of PI3K and PKB inhibits transcriptional regulation by a number of forkhead transcription factors (FoxO1, FoxO3, and FoxO4).[14] PKB phosphorylation of these transcription factors results in their dissociation from their REs and translocation to the cytoplasm. In the absence of IL-2, DNA-bound FoxO3 promotes the expression of p27^{Kip1}.[15] Through the PI3K/PKB-mediated phosphorylation of FoxO3, IL-2 causes the dissociation of FoxO3 from its REs, thereby leading to the attenuation of p27^{Kip1} expression, which ultimately facilitates the activation of cyclin E/CDK2, and passage through the G_1 restriction point. In addition, IL-2-stimulated PI3K/PKB phosphorylation also removes FoxO3 transcriptional repression of cyclin D2 gene expression.[16] Thus, the IL-2 activation of the PI3K/PKB pathway has a major impact on the control of cell cycle progression through the G_1 restriction point by removing the FoxO3 transcriptional activation of the major CKI, p27^{Kip1}, and by removing the FoxO3 transcriptional repression of the cyclin D2 gene.

These data complement other findings from the Cantrell group, which have shown that the PI3K pathway via PKB activates E2F, the transcriptional activating factor that is repressed by the Rb pocket proteins.[17] PKB does not directly phosphorylate Rb. Instead, it appears to function via promoting the expression of cyclin D3, and accelerating the disappearance of p27^{Kip1}, thereby facilitating the cyclin D/CDK phosphorylation of Rb and activating E2F.

All of these PI3K/PKB processes are rapidly reversed when IL-2 becomes depleted through receptor-mediated endocytosis, or if it is withdrawn. Thus, FoxO3 migrates back to the nucleus and rebinds to its REs, thereby promoting expression of p27^{Kip1}, repressing the transcription of cyclin D, thus favoring arrest of the cell in G_1. In addition, hypophosphorylated Rb re-binds E2F, repressing its activity. Furthermore, in the very first experiments with cloned murine IL-2-dependent CTLL cells, it was apparent that removal of IL-2 not only resulted in arrest of the cells in early G_1, but also subsequently led to rapid cell death, a phenomenon that came to be known as cytokine deprivation apoptosis. Actually, this phenomenon facilitated the sensitivity and accuracy of the IL-2 bioassay, in that within 18–24 hours, CTLL cells left without IL-2 promptly die, thereby ensuring a very low background ^{3}H-TdR incorporation and a large signal/background ratio.[18]

The molecular mechanisms responsible for cytokine deprivation apoptosis are mediated in part by the FoxO3 activation of the Bim gene. Bim is a pro-apoptotic BH3 domain-only member of the Bcl-2 family. The Bcl-2 family consists of both pro-and anti-apoptotic molecules that can either homodimerize or heterodimerize to titrate their opposite functions. Thus, either up-regulation of a pro-apoptotic member or down-regulation of an anti-apoptotic member can lead to induction of mitochrondrial apoptosis. In this regard it is noteworthy that PKB induces the expression of the anti-apoptotic molecule Bcl-2 and inactivates the pro-apoptotic molecule Bad. With regard to Bim, the IL-2-activation of the PI3K/PKB pathway down-regulates FoxO3-induced expression of Bim,[15] thereby opposing apoptosis. Accordingly, upon IL-2 withdrawal, FoxO3-mediated reactivation of Bim expression favors apoptosis. It is also of interest that Bim up-regulation upon IL-2 withdrawal follows up-regulation of $p27^{Kip1}$, which is consistent with the observation that apoptosis follows G_0/G_1 arrest.

It is especially noteworthy that the Cip/Kip proteins are intrinsically unstructured, adopting specific tertiary conformations only after binding to other proteins. The crystal structure of the N-terminal and CDK-binding domains of $p27^{Kip1}$ bound to cyclin A-CDK2 revealed that the CKI occludes a substrate interaction domain on the cyclin subunit and inserts itself in the catalytic cleft of the CDK, thereby preventing ATP binding and catalytic activity.[19] However, the binding specificity of the Cip/Kip proteins is modulated by their phosphorylation on distinct residues, and their potency to inhibit cyclin-CDK complexes can be modified by binding to other proteins. In this regard, increasing evidence points to the importance of subcellular localization in the control of the function of p27, and raises the possibility that cytoplasmic p27 may actively contribute to tumorigenesis, independently of its interaction with cyclin-CDK complexes. Thus, mutation of the residues responsible for cyclin/CDK binding, thereby preventing its inhibitory role in cell cycle progression, leads to hyperplasia and spontaneous tumorigenesis in multiple organs, including the lung, retina, pituitary, ovary, adrenals, and spleen.[20,21]

The mitogen activated protein kinase (MAPK) pathway is the other pathway activated via the recruitment of Shc to Y338. It is also know as the Ras/Raf/MEK/Erk pathway. Like the PI3K pathway, the MAPK pathway is also activated via the TCR, so that the role of this pathway in promoting TCR-dependent cell cycle competence versus IL-2-dependent cell cycle progression remains unclear. The guanine nucleotide-binding proteins (H-Ras, K-Ras, N-Ras) were among the first cellular proto-oncogenes discovered. For example, the Harvey murine sarcoma virus (Ha-MSV) originated from the plasma from a leukemic rat that had been inoculated with Moloney murine leukemia virus (Mo-MLV).[22] When injected into newborn mice, this rat-passaged Mo-MLV virus was found to induce the rapid onset of solid tumors (fibrosarcomas), and curiously, erythroblastosis. Subsequently, Ha-MSV was found to contain rat sequences inserted into the MLV genome that encode a 21 kDa molecule that has guanine nucleotide-binding activity.[23]

The TCR was the first extracellular stimulus found to activate p21^{ras},[24] and the MAPK pathway has subsequently been shown to be involved in signaling T cell cycle competence through the activation of the AP-1 transcription factor family, which is one of the transcription factors required for IL-2 gene expression.[25,26] Subsequent experiments by Antanina Zmuidzinas showed that the serine/threonine-specific kinase, Raf-1, which is downstream of p21^{ras}, is also activated by both the TCR and the IL-2R.[27] Raf-1 is the normal cellular homologue of *v-raf*, which like *v-ras* acutely transforms fibroblasts *in vitro* and *in vivo*.[28] Therefore, it is recognized that the MAPK pathway functions to regulate fibroblast proliferation, but its role in signaling normal T cell cycle progression has remained obscure, especially as other cytokines, such as IL-4 and IL-7, can induce T cell cycle progression in the absence of activating MAPK.[29–31] Moreover, as the IL-2R triggers p21^{ras} activation as well,[32] it may be involved in signaling other cellular changes mediated by a cooperative interaction between the TCR and IL-2R in T cells. As noted recently by Doreen Cantrell, "analysis of the human genome reveals the presence of several hundred genes that encode serine/threonine kinases, and the function of most of these kinases in the context of immunocyte signaling remains unknown."[33] Until these issues are resolved, exactly how the MAPK pathway contributes to cytokine-mediated quantal cell cycle progression, and how this pathway is coupled to the digital signals emanating from the Y338 of the IL-2R will remain obscure.

References

1. Friedman, M., Migone, T., Russell, S., and Leonard, W. (1996) Different IL-2 receptor beta chain residues couple to at least two signaling pathways and synergistically mediate interleukin-2-induced proliferation. *Proc. Natl. Acad. Sci. USA* **93**:2077–2082.

2. Lockyer, H., Tran, E., and Nelson, B. (2007) STAT5 is essential for Akt/p70S6 kinase activity during IL-2-induced lymphocyte proliferation. *J. Immunol.* **179**:5301–5308.

3. Zhou, M.-M., Meadows, R., Logan, T., Yoon, H., Wade, W., Ravichandran, K., Burakoff, S., and Fesik, S. (1995) Solution structure of the Shc SH2 domain complexed with a tyrosine-phosphorylated peptide from the T cell receptor. *Proc. Natl. Acad. Sci. USA* **92**:7784–7788.

4. Cantrell, D., Izuierdo, M., Reif, K., and Woodrow, M. (1993) Regulation of PtdIns-3-kinase and the guanine nucleotide binding proteins p21ras during signal transduction by the T cell antigen receptor and the interleukin-2 receptor. *Semin. Immunol.* **5**:319–326.

5. Edinger, A., and Thompson, C. (2002) Akt maintains cell size and survival by increasing mTOR-dependent nutrient uptake. *Mol. Biol. Cell.* **13**:2276–2288.

6. Cornish, G., Sinclair, L., and Cantrell, D. (2006) Differential regulation of T cell growth by IL-2 and IL-15. *Bolld* **108**:600–608.

7. Warberg, O. (1930) *The Metabolism of Tumors.* Arnold Constable. London, UK.

8. Fantin, V., St.-Pierre, J., and Leder, P. (2006) Attenuation of LDH-A expression uncovers a link between glycolysis, mitochondrial physiology, and tumor maintenance. *Cancer Cell* **9**:425–434.

9. Bui, T., and Thompson, C. (2006) Cancer's sweet tooth. *Cancer Cell* **9**:419–420.

10. Sinclair, L., Finlay, D., Feijoo, C., Cornish, G., Gray, A., Ager, A., Okkenhaug, K., Hagenbek, T., Spits, H., and Cantrell, D. (2008) Phosphatidylinositol-3-OH kinase and nutrient sensing mTOR pathways control lymphocyte trafficking. *Nature Immunol.* **9**:513–521.

11. Firpo, E., Koff, A., Solomon, M., and Roberts, J. (1994) Inactivation of a Cdk2 inhibitor during interleukin-2-induced proliferation of human T lymphocytes. *Mol. Cell. Biol.* **14**:44889–44901.

12. Nourse, J., Firpo, E., Flanagan, W., Coats, S., Polyak, K., Lee, M., Massague, J., Crabtree, G., and Roberts, J. (1994) Interleukin-2-mediated elimination of the p27Kip1 cyclin-dependent inhibitor prevented by rapamycin. *Nature* **372**:570–573.

13. Turner, J. (1993) IL-2-dependent induction of G1 cyclins in primary T cells is not blocked by rapamycin or cyclosporin A. *Int. Immunol.* **10**:1199–1209.

14. Dijkers, P., Medema, R., Pals, C., Banerji, L., Thomas, N., Lam, E.-F., Burgerling, B., Raaijmakers, J., Lammers, J.-W., Koenderman, L., *et al.* (2000) Forkhead transcription factor FKHR-L1 modulates cytokine-dependent transcriptional regulation of p27Kip1. *Mol. Cell. Biol.* **20**:9138–9148.

15. Stahl, M., Dijkers, P., Kops, G., Lens, S., Coffer, P., Burgering, B., and Medema, R. (2002) The forkhead transcription factor FoxO regulates transcription of p27Kip1 and bim in response to IL-2. *J. Immunol.* **168**:5024–5031.

16. Schmidt, M., de Mateos, S., van der Horst, A., Klompmaker, R., Kops, G., Lam, E.-F., Burgering, B., and Medema, R. (2002) Cell cycle inhibition by FoxO forkhead transcription factors involves downregulation of cyclin D. *Mol. Cell. Biol.* **22**:7842–7852.

17. Brennan, P., Babbage, J., Burgering, B., Groner, B., Reif, K., and Cantrell, D. (1997) Phosphatidylinositol 3-kinase couples the interleukin-2 receptor to the cell cycle regulator E2F. *Immunity* **7**:679–689.

18. Gillis, S., Ferm, M.M., Ou, W., and Smith, K.A. (1978) T cell growth factor: parameters of production and a quantitative microassay for activity. *J. Immunol.* **120**:2027–2032.

19. Russo, A., Jeffrey, P., Patten, A., Massague, J., and Pavletich, N. (1996) Crystal structure of the p27Kip1 cyclin-dependent-kinase inhibitor bound to the cyclin A-CDK2 complex. *Nature* **325**:325–331.

20. Besson, A., Hwang, H., Cicero, S., Donovan, S., Gurian-West, M., Johnson, D., Clurman, B., Dyer, M., and Roberts, J. (2007) Discovery of an oncogenic activity in p27Kip1 that causes stem cell expansion and a multiple tumor phenotype. *Genes Develop.* **21**:1731–1746.

21. Besson, A., Dowdy, S., and Roberts, J. (2008) CDK inhibitors: cell cycle regulators and beyond. *Develop. Cell* **14**:159–169.

22. Harvey, H. (1964) An unidentified virus which causes the rapid production of tumors in mice. *Nature* **204**:1104–1105.

23. Shih, T., Papageorge, A., Stokes, P., Weeks, M., and Scolnick, E. (1980) Guanine nucleotide-binding and autophosphorylating activities associated with the p21src protein of Harvey murine sarcoma virus. *Nature* **287**:686–691.

24. Downward, J., Graves, J., Warne, P., Rayter, S., and Cantrell, D. (1990) Stimulation of p21ras upon T cell activation. *Nature* **346**:719–723.

25. Garrity, P.A., Chen, D., Rothenberg, E.V., and Wold, B.J. (1994) Interleukin 2 transcription is regulated *in vivo* at the level of coordinated binding of both constitutive and regulated factors. *Mol. Cell. Bio.* **14**:2159–2169.

26. Rothenberg, E.V., and Ward, S.B. (1996) A dynamic assembly of diverse transcription factors integrates activation and cell-type information for interleukin 2 gene regulation. *Proc. Natl. Acad. Sci. USA* **93**:9358–9365.

27. Zmuidzinas, A., Mamon, H.J., Roberts, T.M., and Smith, K.A. (1991) Interleukin 2-triggered Raf-1 expression, phosphorylation, and associated kinase activity increase through G1 and S in CD3-stimulated primary human T cells. *Mol. Cell. Bio.* **11**:2794–2803.

28. Rapp, U., Goldsborough, M., Mark, G., Bonner, T., Groffen, J., Reynolds, F.J., and Stephenson, J. (1983) Structure and biological activity of v-raf, a unique oncogene transduced by a retrovirus. *Proc. Natl. Acad. Sci. USA* **80**:4218–4222.

29. Satoh, T., Nakafuku, M., Miyajima, A., and Kaziro, Y. (1991) Involvement of ras p21 protein in signal-transduction pathways from interleukin-2, interleukin-3,

and granulocyte/macrophage colony-stimulating factor, but not from inter-leukin-4. *Proc. Natl. Acad. Sci. USA* **88**:3314–3318.

30. Crawley, J., Willcocks, J., and Foxwell, B. (1996) Interleukin-7 induces T cell proliferation in the absence of Erk/MAP kinase activity. *E. J. Immunol.* **26**:2717–2723.

31. Welham, M., Duronio, V., and Schrader, J. (1994) Interleukin-4-dependent proliferation dissociates p44erk-1, p42erk-2 and p21ras activation from cell growth. *J. Biol. Chem.* **269**:5865–5873.

32. Izquierdo, M., and Cantrell, D. (1993) Protein tyrosine kinases couple the inter-leukin-2 receptor to p21ras. *E. J. Immunol.* **23**:131–135.

33. Matthews, S., and Cantrell, D. (2006) The role of serine/threonine kinases in T cell activation. *Current Opinion in Immunol.* **18**:314–320.

Chapter 11

The T Cell Antigen Receptor Complex and the Quantal Regulation of the IL-2 and IL-2R Genes

Once it is known that the number of triggered IL-2Rs determines quantal T cell DNA replication and cytokinesis, and thereby both the number of clones responding, as well as the extent of the clonal expansion (i.e. the systemic T cell immune response), the next critical molecular question is what determines the number of IL-2 molecules produced/cell and the number of IL-2Rs expressed by an antigen-activated T cell? Thus, working backward from IL-2 protein molecules to the IL-2 gene, and from the transcription factors and signaling pathways regulating IL-2 gene expression, we know that only antigen-activated T cells express IL-2 genes. After the antigen is cleared, there is no longer a positive drive on the system, and consequently immunologists have been comfortable with the notion that the regulation of the immune response is entirely antigen-dependent, despite the added complexity of the endogenous leukocytotophic hormonal system.

The IL-2 gene promoter[1] has been the most extensively studied of all cytokine gene promoters, and has been found to be regulated by the formation of an enhancesome (a large macromolecular complex of several transcription factors and co-factors) consisting of members of three distinct families of transcription factors, which include activating protein-1 (AP-1), nuclear factor of activated T cells (NF-AT), and Rel.[2-4] Individual transcription factors from these three families cannot bind stably to their REs in the IL-2

87

promoter without all three factors being present simultaneously. If a member of any one of these molecules is prevented from participating, e.g. by immunosuppressive drugs, a marked attenuation of IL-2 gene transcription occurs. In addition, even after the factors have bound to their REs, pharmacological inactivation of any of these three transcription factors extinguishes the binding of all three factors, essentially aborting further IL-2 gene transcription.

In addition, for stable and continuous transcription of the IL-2 gene, co-stimulation via one of the accessory molecules of the CD28 family must be engaged. This phenomenon was first described by Jonathon Lamb and Marc Feldmann using antigenic peptide-activated T cell clones,[5] and later popularized by Marc Jenkins and Ronald Schwartz.[6] Thus, activation of the TCR complex, consisting of an $\alpha\beta$ heterodimer non-covalently linked to the CD3 complex and the accessory molecules CD4 or CD8 (Signal #1) without a CD28 co-stimulatory molecule (Signal #2), was found to lead to an apparent tolerant state *in vitro*, in that the cells could no longer be activated via the TCR and CD28 to produce IL-2 and to proliferate, a cellular fate defined as "anergy." In this case subsequent studies have shown that abortive expression of the IL-2 gene occurs, and the cells are rendered anergic via the activation of a negative regulatory gene program, as well as the degradation of TCR signaling components, such as PLCγ and PKCθ.[7,8] Caspace-3 is one of the genes expressed as a result of TCR/CD3 activation in the absence of co-stimulation.[9] In anergized T cells, upon reactivation of the TCR/CD3 complex caspace-3 associates with the plasma membrane where it is cleaved and inactivates the MAPK pathway, thereby causing a block in TCR signaling of IL-2 gene expression by preventing the participation of AP-1.

The TCR/CD3 complex also activates the p65 subunit of the Rel transcription factor family (a.k.a. Rel A, a.k.a. NF-κB), which initiates IL-2 gene transcription together with AP-1 and NF-AT. However, for sustained IL-2 gene transcription, the c-Rel subunit of the Rel family must also be activated via the co-stimulatory molecule CD28.[10-14] Therefore, two signals are required for full T cell activation because a stable macromolecular enhancesome complex must be formed at the

IL-2 promoter, consisting of AP-1, NF-AT, p65/50 and/or c-Rel. Ellen Rothenberg and co-workers have shown that this is a nonhierarchical, cooperative enhancesome formed by these three families of transcription factors that drives IL-2 gene expression in an all-or-none (quantal) fashion.[3,4]

The quantal regulation of IL-2 gene expression is best illustrated by experiments focused on allelic expression. Ronald Schwartz and co-workers showed that under optimal TCR stimulating conditions, when peptide antigen is in excess and APCs are not limiting, there is biallelic expression of both IL-2 genes.[15] For their experiments, the tremendous heterogeneity of TCRs was circumvented by using transgenic (Tg) T cells that recognize a single peptide from pigeon cytochrome C, and that were made heterozygous (HET) for a gene-targeted mutation of the IL-2 locus. As shown in Fig. 11.1, sorted individual CD4+ T cells from Tg IL-2 WT and Tg IL-2 HET mice were found to have equal frequencies of IL-2 producing cells (approximately 2/3) when assayed after 2.5 days of culture, but the mean amount of IL-2 produced by WT cells was ~2-fold greater than that produced by the HET. In addition, it is noteworthy that the amounts of IL-2 produced by individual cells, as depicted by the individual dots, varied over two orders of magnitude for both WT and HET cells. This broad heterogeneity, even among Tg (and therefore genetically identical) T cells is remarkable and reminiscent of the log-normal distribution of cell cycle times discussed previously. The log-normal distribution of the amounts of IL-2 produced by IL-2 WT versus HET Tg T cells is shown in Fig. 11.2 by IL-2-reactive MoAbs and intracellular staining of cell populations activated via anti-CD3/28. In addition to the two-fold lower MFI of the HET versus the WT cell populations, there is a *log-normal* distribution of the amounts of IL-2 produced by each individual cell within the population. In other experiments when the activating stimuli (i.e. antigenic peptide concentrations) were reduced, there was a progressive decrease in the *proportion* of cells capable of producing IL-2, as well as a decrease in the amounts of IL-2 produced by both the IL-2 WT and HET cells. Accordingly, the magnitude of the IL-2 response is antigen-dependent, but the

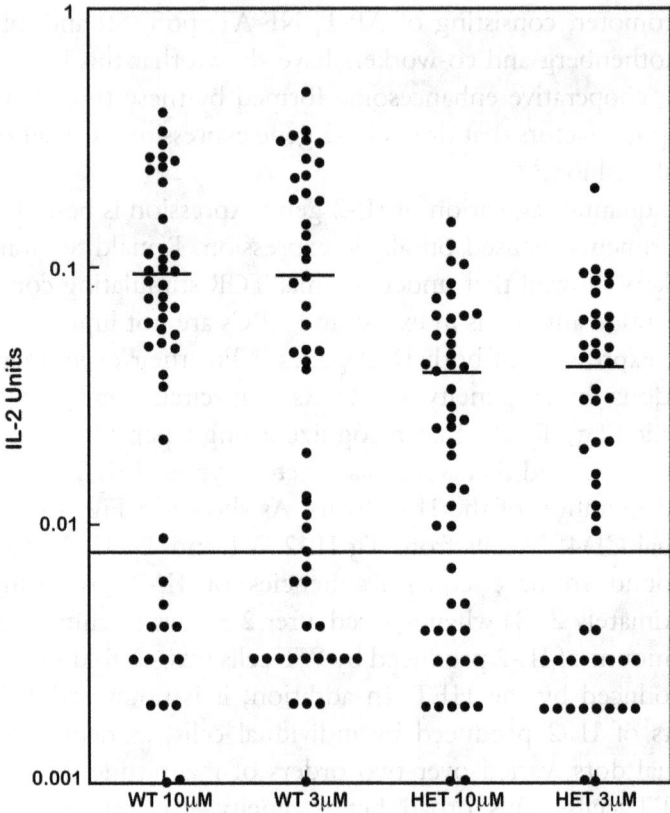

Figure 11.1:   Monoclonal TCR transgenic (Tg) T cell IL-2 production varies in a log-normal fashion. CD4+ T cells from TCR Tg/IL-2 WT, or Tg/IL-2 heterozygotes (HET) were cloned by visualization and cultured for 2.5 days with APCs and 10 μM or 3 μM peptide antigen. Culture supernatants were monitored for [IL-2] via a standard CTLL IL-2 bioassay. Each dot represents the IL-2 produced in a single well. The LLD was 0.008 U/mL, and represented by the straight horizontal line. (From: Chiodetti, L. *et al.* 2000. *Eur. J. Immunol.* **30**:2157–2163.)

antigen concentration still does not account for the variability of the individual cell responses observed.

In this regard, it is important to note that even in a clonal population of T cells, each cell of which has the same affinity for the antigen, there still is a log-normal distribution of the number of TCR/CD3/28 complexes expressed by the individual cells within the

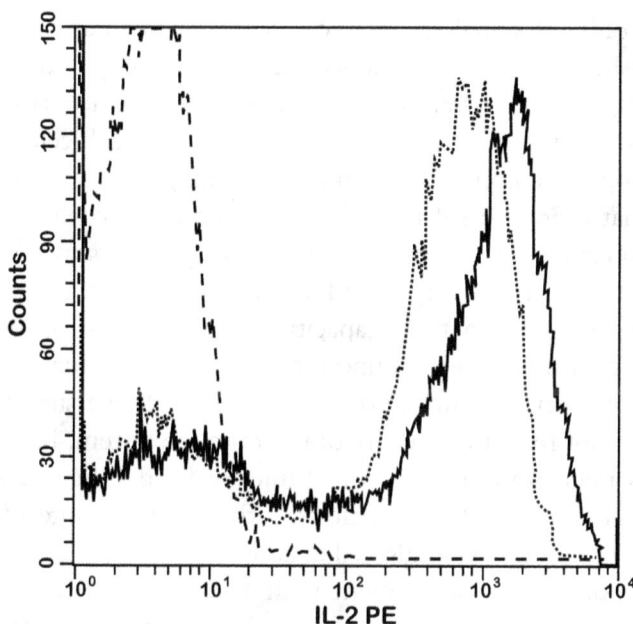

Figure 11.2: **The intracellular accumulation of IL-2 is log-normal within a mono-clonal population.** Tg/IL-2 WT (solid line) and Tg/IL-2 HET (dashed line) cells were activated *in vitro* at optimal anti-CD3 concentrations (20 μg/mL) + anti-CD28, then monitored for intracellular IL-2 expression by flow cytometry. (From: Chiodetti, L. *et al.* 2000. *Eur. J. Immunol.* **30**:2157–2163.)

cell population. Thus, the TCR complex density/cell can differ by two orders of magnitude, with most of the cells distributed about the mean. Detailed studies by Itoh and Germain examining the cytokine response of murine T cell clones to graded antigenic peptide concentrations revealed a hierarchical organization of TCR-dependent response thresholds for elicitation of IL-2 and IFNγ gene expression by individual cells.[16] If one exposes the cloned cells to increasing antigen concentrations, the *proportion* of cells capable of producing detectable cytokines increases. Moreover, there is a greater proportion of cells producing IFNγ versus IL-2, and a higher antigen concentration is necessary for IL-2 production as compared with IFNγ. Thus, rather than a fixed differentiation program, the amounts and kinds of cytokines produced is dependent upon the number and duration of TCR signals generated, according to an apparent activation threshold, which varies among cytokine genes. However, at a given antigen concentration, the decision to produce IL-2 or any other cytokine is a quantal decision on the part of each cell. Thus, the immunological community is now becoming skeptical about the notion that there are subsets of T cells that are fixed lineages, only capable of expressing a limited repertoire of cytokine genes.[17] Instead, investigators are beginning to entertain the idea that T cells may be plastic or malleable in their capacity to produce a diverse array of cytokine genes, depending upon the molecular signals that they receive from their cell surface receptors. Thus, Th1 versus Th2 versus Th17 may not reflect the reality of the immune system, but rather an artificial categorization, based on studies of fixed antigen concentration stimulation of cell populations, rather than a range of antigen densities together with single cell analysis.

These data are in agreement with a series of experiments from Antonio Lanzavecchia's group, which showed that like IL-2R signaling the "*strength*" of the TCR signal is determined by the antigen concentration, the affinity and density of the TCR expressed by an individual cell, and the duration of the antigen/TCR interaction.[18–20]

However, the discovery of the supramolecular activation cluster (SMAC),[21] or the Immunological Synapse,[22] largely superceded these analyses of TCR signal generation. The SMAC or Immunological Synapse is a very large and stable macromolecular complex that forms between a T cell and an APC. On the T cell, the synapse is comprised of a central cluster of TCR/CD4/8 molecules together with co-stimulatory molecules (e.g. CD28) as well as co-inhibitory molecules (e.g. CTLA-4), surrounded by a more peripheral ring of adhesion molecules (e.g. LFA-1). It is noteworthy that most of the TCRs and co-stimulatory molecules expressed by an individual cell participate in the formation of the "central synapse," so that there is a mean of $\sim 10^4$ TCRs focused in this one area. However, because of the log-normal distribution of TCRs among cells within a given population, some cells will express only $\sim 10^3$ TCRs while others could express as many as 10^5 TCRs. It goes without saying that this 100-fold difference in TCR complexes could very well translate into a major difference in the "strength" of signals imparted to the cell interior.

The opposing APC area that abuts the TCR synapse is comprised of a central cluster of antigenic peptide bound to MHC-encoded molecules (pMHC), together with the ligands (termed B7 family molecules) that are capable of stimulating the co-stimulatory (CD28) and co-inhibitory molecules (CTLA-4) expressed by the T cell. Surrounding this central cluster are the ligands capable of binding with the adhesion molecules expressed on the T cells.

The function of this SMAC/immunological synapse has been under intense experimental scrutiny for the past decade. Because of the arrangement of the central and peripheral molecules, this macromolecular complex was assumed early on to be involved in signaling. However, it was found that T cell receptor signaling is initiated *before* the formation of the synapse.[23] Moreover, deletion of the adapter molecule CD2-AP prevents cells from forming a mature T-B synapse, yet the CD4+ T cells are able to produce even greater quantities of IL-2 and proliferate to a greater extent than normal cells.[24] Thus, the formation of the synapse may have a function independent from signaling.

In this regard, recent data from Ellis Reinherz's group indicates that the TCR is an anisotropic mechanoreceptor, which converts mechanical energy into a biochemical signal.[25] According to their results, an external torque on the TCR complex quaternary structure following pMHC ligation during T cell scanning of an APC serves as the source of energy for directional force. Since the total force applied to the T cell surface is produced during movement of the T cell membrane across the membrane of the APC, ligation of several TCRs by specific cognate pMHCs will exert a greater physical force on each individual TCR than multiple non-cognate, nonspecific TCR-pMHC interactions. Thus, specificity and sensitivity are built into the TCR mechanoreceptor function. Moreover, because this non-equilibrium signaling mechanism is anisotropic, isotropic equilibrium constants may be insufficient to describe mechanoreceptor signaling kinetics.

The most logical explanation for the heterogeneity of IL-2 production in response to stimulation by varying pMHC concentrations, is that only cells with the highest density of TCRs are capable of responding to the lowest concentration of peptide. Moreover, at TCR-saturating pMHC concentrations, the cells with the highest TCR densities would be expected to produce the greatest amounts of IL-2. Thus, the number of pMHC/TCR interactions over time determines the number of IL-2 molecules produced. This idea has been tested *in vivo* by the construction of TCR transgenic mice with drug-regulated expression of TCRs.[26] The pMHC/TCR system is organized to be very sensitive to low antigen concentrations by the expression of very high densities of TCRs. Thus, by adjusting the mean TCR density to ~20,000 sites/cell, very low antigenic peptide doses administered intravenously (IV) maximally activate IL-2 production and proliferation. Reducing the mean number of TCRs/cell to <1000 results in higher peptide concentrations which are necessary to promote a detectable response.

Other studies have made it possible to count the exact number of pMHC ligands that a T cell encounters on an APC, monitoring the interaction by quantifying increases in intracellular calcium concentrations. Only one pMHC ligand can trigger a transient rise in intracellular calcium concentration, while 10 pMHC ligands/APC are sufficient to sustain an increased intracellular calcium concentration. Moreover, at least 10 hours of stable T cell/APC interaction is necessary to promote optimal IL-2 gene expression. Accordingly, just like the number of IL-2/IL-2R interactions necessary to trigger expression of a crucial number of cyclin D2 molecules and cause the disappearance of p27^{Kip1} molecules to produce a quantal decision to proceed to S-phase, serial pMHC triggering of the TCR is necessary to activate and sustain maximal IL-2 gene expression.[27] However, in contrast to the IL-2/IL-2R system, where the ligand is in excess while the receptor is limiting, the pMHC/TCR system is the reverse. There is a great excess of TCRs, which can be serially triggered by very few agonist ligand pMHC molecules, <10 molecules/cell. Thus, by virtue of a vast excess of TCRs, antigen recognition is very sensitive. However, a long time interval is necessary to trigger a quantal cellular response, indicating that a counting mechanism somehow transfers the information generated at the cell surface to the nucleus and the IL-2 gene.

Extrapolating to the systemic immune response, it becomes obvious that the number of naive T cells that are productively triggered, and the duration that IL-2 and other cytokines are expressed, will be reflected ultimately by the number and size of the expanded clones of cells.[28] This will determine whether there is a detectable systemic immune response to a particular pMHC, whether it is non-self or self in origin. When the pMHC antigen concentration increases, T cells with lower TCR affinities and/or densities will be brought into play, thereby resulting in a greater systemic immune response. Because self-peptides and foreign peptides are indistinguishable, both being peptides, the number of pMHC molecules presented by an APC, and the densities and affinities of the TCRs that can react will determine whether a systemic immune response ensues. It follows that *the number of self-peptides present on APCs in the periphery must normally be low.* In addition, the number of T cell clones with the potential to recognize a particular self-pMHC with a reasonably high affinity must also be low. Otherwise, autoimmunity would be much more commonplace than it is.

References

1. Holbrook, N.J., Smith, K.A., Fornace, A.J., Jr., Comeau, C.M., Wiskocil, R.L., and Crabtree, G.R. (1984) T-cell growth factor: complete nucleotide sequence and organization of the gene in normal and malignant cells. *Proc. Natl. Acad. Sci. USA* **81**:1634–1638.
2. Shaw, J.P., Utz, P.J., Durand, D.B., Toole, J.J., Emmel, E.A., and Crabtree, G.R. (1988) Identification of a putative regulator of early T cell activation genes. *Science* **241**:202–205.
3. Garrity, P.A., Chen, D., Rothenberg, E.V., and Wold, B.J. (1994) Interleukin 2 transcription is regulated *in vivo* at the level of coordinated binding of both constitutive and regulated factors. *Mol. Cell. Bio.* **14**:2159–2169.
4. Rothenberg, E.V., and Ward, S.B. (1996) A dynamic assembly of diverse transcription factors integrates activation and cell-type information for interleukin 2 gene regulation. *Proc. Natl. Acad. Sci. USA* **93**:9358–9365.
5. Lamb, J., Skidmore, B., Green, N., Chiller, J. and Feldmann, M. (1983) Induction of tolerance in influenza virus-immune T lymphocyte clones with synthetic peptides of influenza hemagglutinin. *J. Exp. Med.* **157**:1434–1447.
6. Jenkins, M., and Schwartz, R. (1987) Antigen presentation by chemically modified splenocytes induces antigen-specific T cell unresponsiveness *in vitro* and *in vivo. J. Exp. Med.* **165**:302–319.

7. Macian, F., Garcia-Cozar, F., Im, S.H., Horton, H.F., Byrne, M.C., and Rao, A. (2002) Transcriptional mechanisms underlying lymphocyte tolerance. *Cell* 109:719–731.

8. Heissmeyer, V., Macian, F., Im, S.H., Varma, R., Feske, S., Venuprasad, K., Gu, H., Liu, Y.C., Dustin, M.L., and Rao, A. (2004) Calcineurin imposes T cell unresponsiveness through targeted proteolysis of signaling proteins. *Nat. Immunol.* 5:255–265.

9. Puga, I., Rao, A., and Macian, F. (2008) Targeted cleavage of signaling proteins by caspace-3 inhibits T cell receptor signaling in anergic T cells. *Immunity* 29:194–204.

10. Thompson, C.B., Lindsten, T., Ledbetter, J.A., Kunkel, S.L., Young, H.A., Emerson, S.G., Leiden, J.M., and June, C.H. (1989) CD28 activation pathway regulates the production of multiple T-cell-derived lymphokines/cytokines. *Proc. Natl. Acad. Sci. USA* 86:1333–1337.

11. Fraser, J.D., Irving, B.A., Crabtree, G.R., and Weiss, A. (1991) Regulation of interleukin-2 gene enhancer activity by the T cell accessory molecule CD28. *Science* 251:313–316.

12. Verweij, C.L., Geerts, M., and Aarden, L.A. (1991) Activation of interleukin 2 gene transcription via the T cell surface molecule CD28 is mediated through an Nf-kB-like response element. *The Journal of Biochemistry* 266:14179–14182.

13. Ghosh, P., Tan, T.-h., Rice, N.R., Sica, A., and Young, H.A. (1993) The interleukin 2 CD28-responsive complex contains at least three members of the NF-kB family: c-Rel, p50, and p65. *Proc. of the Natl. Acad. Sci. USA* 90:1696–1700.

14. Liou, H., Jin, Z., Tumang, J., Andjelic, S., Smith, K., and Liou, M. (1999) c-Rel is crucial for lymphocyte proliferation but dispensable for T cell effector function. *Int. Immunol.* 11:361–371.

15. Chiodetti, L., Barber, D., and Schwartz, R. (2000) Biallelic expression of the IL-2 locus under optimal stimulation conditions. *Eur. J. Immunol.* 30:2157–2163.

16. Itoh, Y., and Germain, R.N. (1997) Single cell analysis reveals regulated hierarchical T cell antigen receptor signaling thresholds and intraclonal heterogeneity for individual cytokine responses of CD4$^+$ T cells. *J. Exp. Med.* 186:757–766.

17. O'Shea, J., Hunter, C., and Germain, R. (2008) T cell heterogeneity: firmly fixed, predominantly plastic or merely malleable? *Nat. Immunol.* 9:450–453.

18. Viola, A., and Lanzavecchia, A. (1996) T cell activation determined by T cell receptor number and tunable thresholds. *Science* 273:104–106.

19. Valitutti, S., Muller, S., Dessing, M., and Lanzavecchia, A. (1996) Different responses are elicited in cytotoxic T lymphocytes by different levels of T cell receptor occupancy. *J. Exp. Med.* 183:1917–1921.

20. Iezzi, G., Karjalainen, K., and Lanzavecchia, A. (1998) The duration of antigenic stimulation determines the fate of naive and effector T cells. *Immunity* 8:89–95.

21. Monks, C., Freiberg, B., Kupfer, H., Sciaky, N. and Kupfer, A. (1998) Three dimensional segregation of supramolecular activation clusters in T cells. *Nature* **395**.

22. Grakoui, A., Bromley, S.K., Sumen, C., Davis, M.M., Shaw, A.S., Allen, P.M., and Dustin, M.L. (1999) The immunological synapse: a molecular machine controlling T cell activation. *Science* **285**:221–227.

23. Lee, K.-H., Holdorf, A.D., Dustin, M.L., Chan, A.C., Allen, P.M., and Shaw, A.S. (2002) T cell receptor signaling precedes immunological synapse formation. *Science* **295**:1539–1542.

24. Lee, K.-H., Dinner, A., Tu, C., Campi, G., Raychaudhuri, S., Varma, R., Sims, T., Burack, W., Wu, H., Wang, J., *et al.* (2003) The immunological synapse balances T cell receptor signaling and degradation. *Science* **302**.

25. Kim, S., Takeuchi, K., Sun, Z.-Y., Touma, M., Castro, C., Fahmy, A., Lang, M., Wagner, G., and Reinherz, E. (2009) The alpha/beta T cell receptor is an anisotropic mechanoreceptor. *J. Biol. Chem.* **284**:31028–31037.

26. Labrecque, N., Whitfield, L.S., Obst, R., Waltzinger, C., Benoist, C., and Mathis, D. (2001) How much TCR does a T cell need? *Immunity* **15**:71–82.

27. Valitutti, S., Muller, S., Cella, M., Padovan, E., and Lanzavecchia, A. (1995) Serial triggering of many T-cell receptors by a few peptide MHC complexes. *Nature* **375**:148–151.

28. Smith, K. (2006) The quantal theory of immunity. *Cell. Res.* **16**:11–19.

Chapter 12

Digital Signaling via the T Cell Antigen Receptor Complex

Given that three families of transcription factors must be activated in a coordinated and quantal fashion by the TCR complex, the question arises as to how this digital response on the part of the cell is created. Various theories have been proposed to account for the biophysics and biochemistry of TCR complex signaling, including induced allosteric changes in the TCR,[1] the duration of the pMHC/TCR complex interaction,[2] kinetic proofreading,[3] and membrane segregation of signaling proteins (reviewed in Ref. 4). However, as detailed by Altan-Bonnet and co-workers, none of these theories explain the three characteristics of TCR activation, speed, sensitivity and specificity, which they have termed the S^3 characteristics of TCR activation. Thus, theoretical models that explain TCR signaling must simultaneously account for the exquisite specificity of pMHC discrimination (i.e. agonist peptides versus non-agonist peptides), the high sensitivity of activation (i.e. only a few agonistic pMHC complexes trigger), and the speed of the biochemical response (i.e. calcium or ppERK elevations have been found to occur within seconds of TCR occupancy). The mechanoreceptor as described by the Reinherz group provides for a new notion of how the TCR itself may signal to the interior with both specificity and sensitivity, but we must now begin to discern how the signal is propagated via the cytoplasmic signaling pathways.

Ronald Germain's group first reported[5] that non-agonist pMHC ligands predominantly trigger a negative feedback loop in T cells leading to rapid recruitment of the tyrosine phosphatase SHP-1, which desensitizes the TCR by inactivation of the Lck tyrosine kinase. By comparison, they found that agonistic ligands activate a positive feedback circuit involving Lck modification by ppERK, preventing SHP-1 recruitment, and allowing the persistent signaling necessary for gene activation. Altan-Bonnet and Germain then improved on this concept by proposing a model based upon a classic proofreading scheme with differential feedbacks to account for the S^3 characteristics of T cell activation.[6] They constructed a new mathematical model of proximal TCR-dependent signaling, which illustrates that competition between a digital positive feedback based on ppERK activity and an analogue negative feedback involving SHP-1 is critical for defining a sharp ligand-discrimination non-linear (i.e. digital) threshold, and accounting for a swift, sensitive and specific response.

To obtain experimental data testing their model, they chose to examine the MAPK pathway, monitored at the single cell level via flow cytometry by the dual phosphorylation response of ERK (ppERK) by TCR transgenic CD8[+] T cells stimulated with specific peptide-pulsed APCs. They found that the dual phosphorylation of this kinase can be detected when as few as a mean of 10 antigenic peptides/APC are presented. Moreover, there is a symmetrically sigmoid log/dose-response curve for ERK activation of a population of TCR-Tg T cells as shown in Fig. 12.1. It is noteworthy that at the population level, this log/dose-response curve for ERK activation yields an average threshold for the peak response monitored after three minutes to be 24 peptides/APC, with the absolute threshold to trigger the phosphorylation of 100,000 ERK molecules of 8 peptides/APC. These studies also revealed a previously unappreciated aspect of the T cell ERK-signaling response: after three minutes of contact with APCs, the pattern of ppERK expression is strictly bimodal, i.e. the ppERK response of T cells is essentially digital (i.e. quantal). The bimodal distribution could be fitted as a sum of two discrete log-normal distributions, indicating that the individual cell ppERK response is "switch-like," with a nearly infinite Hill coefficient. It is also noteworthy that ppERK after three minutes of activation correlated with downstream gene expression, monitored by CD69 up-regulation, IFNγ production, or cytotoxic activity.

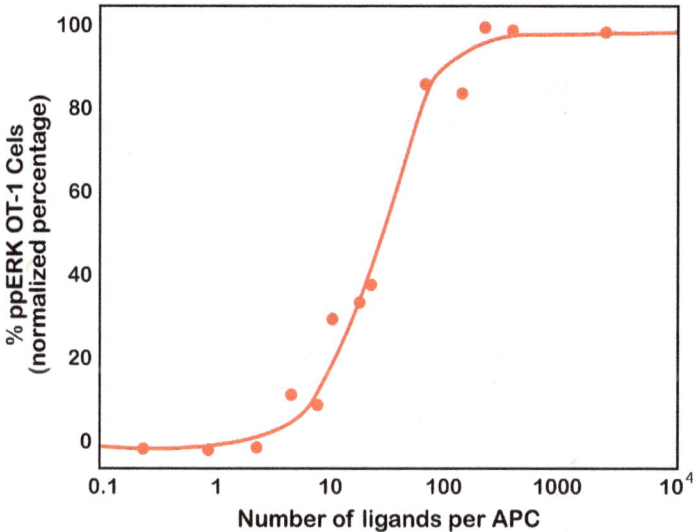

Figure 12.1: A very small number of peptides/APC triggers a maximal ppERK response. Experimental ppERK response of naive TCR transgenic T cells activated for three minutes with peptide-pulsed APCs, plotted as the % of responding cells. The Hill coefficient measured for this dose-response is 1.9 ± 0.1 (n = 3). The threshold for activation (midpoint) is 24 ± 4 peptides on each APC. (From: Altan-Bonnet, G., and Germain, R. 2005. *PLoS Biol.* **3**:e356.)

To explain these results, Altan-Bonnet and Germain proposed a model for TCR signaling composed of a rapid-onset, analogue SHP-1-mediated negative feedback and a slower digital ERK-1-dependent positive feedback modulating the triggering threshold. Several predictions of this model were tested and confirmed by experiment. In particular, an increase in SHP-1 in T cells results in a decreased response to a pMHC ligand. Thus, their results combining computation and experiment indicate that T cell antigen discrimination is regulated by the dynamics of competing feedback loops that control a high-gain digital amplifier, which is dependent upon the intracellular concentration of key enzymes. In addition, given the log-normal distribution of TCRs, even in a clonal population, it is tempting to speculate that the number of TCR complexes/cell and the number of SHP-1 molecules/cell, as well as ERK molecules/cell, are coordinately determined.

In this regard, Altan-Bonnet's group recently examined the effects of the variability of molecules involved in transmitting T cell antigen recognition to the cell interior.[7] First, they theoretically tested how the "natural" variability in the number of signaling components within T cells might affect their response. Using their MAPK activation model, it was predicted that ERK-1 would be a non-critical regulator because its average expression level is already in excess, so that variation within the physiological range would not affect the T cell response. Second, the CD8 co-receptor was predicted by the model to be an analogue regulator, so that higher expression levels of CD8 would considerably decrease the minimum numbers of pMHC ligands required for activation, without affecting the percentage of cells that can respond to the maximum dose of ligands. By comparison, the model predicted that SHP-1 would be a digital regulator. Thus, as SHP-1 levels increase, an increasing fraction of cells becomes unresponsive to any physiologically relevant concentration of pMHC ligands, whereas other cells remain as sensitive as cells expressing low levels of SHP-1.

To test these predictions, the various proteins were quantified in individual cells using MoAbs and flow cytometry. The distributions of expression of CD8, ERK-1, and SHP-1 were log-normal, with coefficients of variations equal to 0.32, 0.56, and 0.62 respectively. Because of these log variations in protein expression, it was considered likely that even though the cells were clonally derived, the cells could not respond identically to antigen stimulation. This was examined via flow cytometry, quantifying in single cells the various proteins and the response of cellular ppERK after five minutes according to increasing pMHC concentrations. Consistent with the predictions of the model, variations in the levels of CD8, ERK-1, and SHP-1 had substantially different effects on the proportion of ppERK-containing cells. CD8 functioned as a positive analogue regulator, such that cells with varying levels of CD8 displayed different sensitivities to pMHC concentrations as manifest by their EC_{50} but had a nearly equal P_{max}. Thus, a six-fold difference in membrane CD8 expression was associated with a 100-fold change in pMHC EC_{50}. By comparison, SHP-1 acted as a negative digital regulator: Cells with low SHP-1 levels had higher P_{max}, but similar EC_{50}, while variation in ERK-1 expression changed neither P_{max} nor EC_{50}, as predicted by the model. In other computer-modeled responses, it was shown that co-regulation of CD8 and SHP-1 expression results in less variation in ppERK signaling than if co-regulation was not functioning. Thus, co-regulation decreases the number of hyper-responsive cells, which could limit the risk of self-responsiveness and autoimmune activation.

The importance of these theoretical and experimental assessments of TCR signaling lie in revealing how single cell analysis can be so informative, especially when signaling molecules are accurately quantified and related with the magnitude of the biological response, in this case ppERK, which also must be quantified carefully. Accordingly, it can readily be seen how semi-quantitative Western blots of activated cell populations are so unrevealing, as well as potentially misleading in the assessment of pathways that signal biological responses. In this regard, since the number of CD8 molecules/cell are responsible for an analogue or graded ppERK response, and because the number of CD8 molecules/cell equals the number of TCR/CD3 complexes/cell, one must account for the quantal expression of IL-2 by graded antigen-activated T cells.

Many of the experiments focused on TCR signaling have monitored increases in intracellular calcium which occur within seconds, or the phosphorylation of CD3 signaling components and/or enzymes, such as ppERK, which occur within minutes. However, ultimately the most relevant response is the expression of the genes responsible for regulating the expression of effector functions, such as cytokines and cytolytic molecules, which often occurs over several hours. Just as in the quantal proliferative response to IL-2, with the advent of the capacity to detect the accumulation of intracellular cytokines by flow cytometry, it became obvious that single cells also express detectable IL-2 and other cytokines in a log-normal heterogeneous, quantal fashion. To explain binary (quantal) IL-2 expression, early reports suggested that the thresholds for IL-2 expression by individual cells are set by the concentrations of transcriptionally active NF-AT and Rel, as well as the cooperation between them, with a stochastic DNA binding of those transcription factors due to a limited availability.[8,9]

Accordingly, this hypothesis is reminiscent of the fallacious probabilistic theory proposed almost 40 years ago to explain the variability of cell cycle times.[10,11]

By comparison, in a recent report, Podtschaske and co-workers show that NF-AT acts as a binary molecular switch. All NF-AT molecules are dephosphorylated in the cytoplasm instantaneously by the phosphatase calcineurin, such that the entire cellular complement of NF-AT molecules translocates into the nucleus simultaneously, thereby activating quantal IL-2 gene expression.[12] Thus, graded concentrations of TCR activation change the *proportion* of IL-2-expressing cells monitored five hours after initiation, but not the amount of IL-2 expressed/cell, as detected by the mean fluorescent intensity (MFI) of anti-IL-2 MoAb staining. The now familiar log-normal variability of individual cell IL-2 expression is also noteworthy, such that some cells express as much as 100-fold greater amounts of IL-2 than others, with most cells distributed close to the mean.

The binary decision for expression of IL-2 was found to be calcium concentration-dependent, and to be dependent upon translocation of dephosphorylated NF-AT to the nucleus. By comparison, in response to graded TCR signaling nuclear translocation of NF-κB is also graded or analogue, while NF-AT nuclear translocation is binary or digital. Moreover, IFNγ, which is also regulated by NF-AT, was expressed in a binary fashion, while CD69 expression, which is not regulated by NF-AT, is expressed in a graded fashion. These findings indicate that the dephosphorylation of NF-AT creates a molecular switch for the binary (quantal) control of IL-2 and IFNγ gene expression.

Mathematical modeling of NF-AT dephosphorylation revealed that of the 13 phosphoserines, the Hill coefficient increases as the number of serines are dephosphorylated, with the steepness of the symmetrically sigmoid log-dose-response curves increasing correspondingly as shown in Fig. 12.2A. By comparison, if NF-AT could be activated through dephosphorylation of a single serine ($n = 1$), a hyperbolic response curve for the nuclear localization of NF-AT would result. Thus, for non-cooperative NF-AT activation, the predicted IL-2 expression pattern is graded, while for high

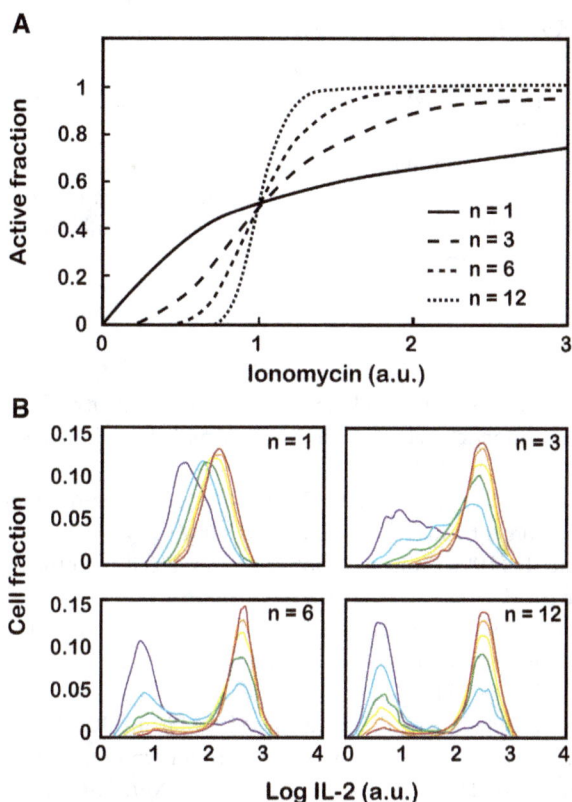

Figure 12.2:   Mathematical modeling of NF-AT-induced IL-2 expression. A. Response curves of NF-AT activity versus activating signal (ionomycin) are shown modeled as a Hill function. Regulation of NF-AT activity by a single phosphorylation site would result in a hyperbolic dose-response (Hill coefficient $n = 1$), while cooperative dephosphorylation of multiple sites would give rise to sigmoid responses ($n > 1$). **B.** Simulated patterns of IL-2 expression. The ionomycin stimulus increases from blue to red in steps of 0.125, running from 0.125 to 0.75 for $n = 1$, and steps of 0.5, running from 0.5 to 3.0 for $n > 1$. With increasing steepness of the response curve, IL-2 expression changes from gradual to binary. (Redrawn from: Podschaske, M. et al. 2007. *PLoS ONE.* **2**:e935.)

cooperativity ($n = 12$) the pattern is binary, as shown in Fig. 21B. Actually, for $n = 6$ and larger, the binary pattern is fully developed and closely matches the experimentally observed pattern for single cell IL-2 expression.

Thus, according to Podtschaske and co-workers:

"The identified role of NF-AT dephosphorylation is not just a threshold. It is rather an all-or-none (i.e. quantal) decision to take the step to (IL-2) productive T cell activation .... By translating the (analogue) strength (amounts?) of antigenic T cell stimulation into the *frequency* of cytokine-producing T cells, the NF-AT switch is a general hub for productive adaptive immune responses."

References

1. Rojo, J., and Janeway Jr., C. (1988) The biological activity of anti-T cell receptor V region monoclonal antibodies is determined by the epitope recognized. *J. Immunol.* **140**:1081–1088.
2. Gascoigne, N., Zal, T., and Alam, S. (2001) T cell receptor binding kinetics in T cell development and activation. *Expert Rev. Mol. Med.* **1**:17.
3. McKeithan, T. (1995) Kinetic proofreading in T cell receptor signal transduction. *Proc. Natl. Acad. Sci. USA* **92**:5042–5046.
4. Feinerman, O., Germain, R., and Altan-Bonnet, G. (2008) Quantitative challenges in understanding ligand discrimination by alpha/beta T cells. *Mol. Immunol.* **45**:619–631.
5. Stefanova, I., Hemmer, B., Vergelli, M., Martin, R., Biddison, W., and Germain, R. (2003) TCR ligand discrimination is enforced by competing ERK positive and SHP-1 negative feedback pathways. *Nat. Immunol.* **4**:248–254.
6. Altan-Bonnet, G., and Germain, R.N. (2005) Modeling T cell antigen discrimination based on feedback control of digital ERK responses. *PLoS Biology* **3**:e356.
7. Feinerman, O., Veiga, J., Dorfman, J., Germain, R., and Altan-Bonnet, G. (2008) Variability and robustness in T cell activation from regulated heterogeneity in protein levels. *Science* **321**:1081–1084.
8. Fiering, S., Northrop, J., Nolan, G., Mattila, P., Crabtree, G., and Herzenberg, L. (1990) Single cell assay of a transcription factor reveals a threshold in transcription activated by signals emanating from the T cell receptor. *Genes Dev.* **4**:1823–1834.
9. Pirone, J., and Elston, T. (2004) Fluctuations in transcription factor binding can explain the graded and binary responses observed in inducible gene expression. *J. Theor. Biol.* **226**:111–121.
10. Burns, V., and Tannock, I. (1970) On the existance of a G0 phase in the cell cycle. *Cell Tissue Kinetics* **3**:321–333.

11. Smith, J., and Martin, L. (1973) Do cells cycle? *Proc. Natl. Acad. Sci. USA* **70**:1263–1269.
12. Podtschaske, M., Benary, U., Zwinger, S., Hofer, T., Radbruch, A., and Baumgrass, R. (2007) Digital NFATc2 activation per cell transforms graded T cell receptor activation into an all-or-none IL2 expression. *PLoS ONE* **2**:e935.

Chapter 13

Negative Feedback Regulation of T Cell Antigen Receptor Complex Signaling — Attenuation of IL-2 Gene Expression

Once it was possible to quantify IL-2 concentrations via the bioassay, it was found that upon mitogenic activation, IL-2 could be first detected within six hours, gradually rising to a peak after 48 hours, followed by a more rapid decline to undetectable levels by 96 hours. As already discussed in Chap. 8, one of the explanations for the decline in detectable IL-2 is the cellular metabolism of IL-2 by IL-2R⁺ cells. However, IL-2 appeared to be different than other cytokines expressed upon TCR signaling, e.g. IFNγ, which rises to a plateau several hours after mitogenic activation and does not decline as the cells in culture proliferate. Early on, we hypothesized that perhaps there was a classic negative feedback regulation of IL-2 at play. We looked for a soluble negative regulator that could function to inhibit IL-2 activity in the bioassay, but could find no evidence for one. Even so, the notion that there must be a system to negatively regulate the IL-2-promoted proliferative expansion of antigen-activated T cell clones made biological (and endocrinological) sense, especially as it became obvious that the magnitude of the systemic immune response is controlled by the endogenous IL-2/IL-2R leukocytotrophic hormonal system,[1,2] instead of antigen *per se*. Actually, the existence of a leukocytotrophic hormonal system that regulates the immune response was first proposed by Sir Peter Medawar in 1973,[3] and formally shown to exist by the discovery of the IL-2 molecule and

IL-2Rs 10 years later.[4,5] However, one hallmark of an internally regulated hormonal system is negative feedback control, and this property of an endocrine system was lacking.

To search for putative negative regulators of T cell activation, it was first necessary to delineate the molecular mechanisms responsible for TCR/CD3 complex signaling as detailed in the last chapter, as well as the role of the co-stimulatory molecule CD28. Thus, the stabilization of IL-2 and other cytokine mRNA transcripts mediated by CD28 signaling,[6] and the three- to four-fold enhancement of the rate of IL-2 transcription,[7] together account for delivering the "second signal" of T cell activation.[8,9] In this regard, it is important to emphasize that truly productive T cell activation requires the proliferative signal delivered by the IL-2/IL-2R interaction, which therefore becomes the "third signal" of T cell activation.

The cytotoxic T cell late antigen #4 (CTLA-4) was the first molecule to be discovered that appeared to negatively regulate TCR signaling.[10] First identified in 1987 by screening for T cell-specific genes using the T/B-cell subtractive cDNA hybridization, CTLA-4 appears "Late" after TCR activation, peaking after 48–72 hours. As CTLA-4 is ~30% homologous to CD28, it was expected that it would also serve a co-stimulatory role. However, CTLA-4-blocking MoAbs were found to actually enhance anti-CD3-stimulated T cell proliferation, instead of the anticipated suppression.[11] Subsequently, deletion of the CTLA-4 genes resulted in the surprising finding of massive polyclonal T cell proliferation and rapidly fatal multi-organ tissue destruction by three to four weeks of age, thereby indicating that CTLA-4 might normally act as a negative feedback molecule.[12,13] The lymphoproliferative syndrome exhibited by these mice was profound, in that the thymuses were markedly abnormal, with a five-fold increase in immature double-negative thymocytes, and a 10-fold diminution of double-positive thymocytes, thereby suggesting a block in T cell maturation. By comparison, in the periphery, there was massive lymphadenopathy and splenomegaly, with a 10-fold increase in cellularity, and a relative increase in T cells versus B cells. Moreover, there was a marked increase in T cell surface "activation markers" such as CD69, CD25, and CD44, along with a lymphoblast morphology.

Mononuclear cells infiltrate many organs, resulting in especially prominent destructive myocarditis, and a severe pancreatitis characterized by an intense mononuclear cell infiltrate with destruction of both endocrine and exocrine components. Immunohistochemical analysis indicated that the mononuclear cells are predominantly T cells and macrophages, with minimal contribution by B cells. The lungs, liver, and salivary glands also had mononuclear cell infiltrates, while in sharp contrast the kidney and thyroid glands exhibit no inflammatory changes. Consistent with the conclusion that these cells are antigen-activated *in vivo*, upon short-term culture *in vitro*, the CD4[+] T cells from the CTLA-4 (-/-) released cytokines such as GM-CSF spontaneously, and undergo spontaneous proliferation.

In the very first paper that described the discovery and cloning of CTLA-4 cDNA, it was shown that thymocytes only express CTLA-4 mRNA when activated by both Con-A and IL-2,[10] and subsequent studies showed that CTLA-4 expression is dependent upon TCR and CD28 activation, i.e. signals #1 and #2. However, because these two signals result in maximal IL-2 and IL-2R expression, it was not clear whether TCR/CD28 signals CTLA-4 expression *de novo*, or whether IL-2R signals (i.e. signal #3) also play a role. Because CD28 and CTLA-4 are both members of the Ig superfamily, it was natural to assume that they regulated one another reciprocally. However, CD28-KO mouse T cells were found to have deficient CTLA-4 expression that could be restored by IL-2 supplementation alone.[11] Moreover, IL-2-KO mouse T cells, which have intact TCR and CD28 signaling, also were found to have deficient CTLA-4 expression; CTLA-4 levels in TCR/CD28-activated IL-2-KO CD4[+] T cells were only ~50% of those in WT mice, and completely absent from IL-2-KO CD8[+] T cells; both were restored to normal levels by exogenous IL-2.[14] Furthermore, the effect is IL-2-specific, in that IL-4, IL-6, IL-7, and IL-12 cannot substitute for IL-2. Accordingly, IL-2 promotes CTLA-4 expression by TCR/CD28-activated T cells, so that it is the first example of an IL-2-dependent negative feedback loop, which functions to suppress T cell activation by inhibiting TCR signaling of IL-2 gene expression. Accordingly, a deficiency of either IL-2 or CTLA-4 would be expected to lead to a hyperactive T cell state, which would predispose to the loss of self–non-self recognition and autoimmunity.

Another member of the Ig superfamily found to have co-inhibitory activity is the programmed death-1 (PD-1) molecule, first described in 1992.[15] Like CTLA-4, PD-1 is not expressed by resting lymphocytes, but appears on T cells within 24 hours of stimulation via the TCR/CD3 complex and becomes maximally expressed on the majority (>80%) of both CD4+ and CD8+ T cells after five days of culture.[16,17] It is noteworthy that in contrast to CTLA-4 expression, which is restricted to activated T cells, PD-1 is also expressed by B cells stimulated via the B cell antigen receptor (BCR) and by activated macrophages. Like targeted disruption of the CTLA-4 gene, deletion of the PD-1 genes leads to a polyclonal lymphoproliferative syndrome, but it is much milder and of slower onset.[18] For example, six-week-old PD-1 (-/-) mice exhibit only a doubling of spleen size compared with WT mice, with a 60% increase in B cells, a 40% increase in T cells, and maintenance of a normal CD4+/CD8+ ratio. By comparison, there is almost a 400% increase in the number of myeloid cells. Moreover, widespread mononuclear cell infiltration of multiple organs such as occurs in CTLA-4 (-/-) mice does not occur in PD-1 (-/-) mice. Instead, distinct autoimmune syndromes occur in different mouse strains when the PD-1 genes are deleted, such as a lupus-like disease in C57Bl/6 mice, with glomerulonephritis and arthritis,[19] and dilated cardiomyopathy in BALB/c mice, with the generation of cardiomyocyte-reactive antibodies.[20]

Blockade of the PD-1 negative regulatory pathway has also been found to accelerate autoimmune diseases in susceptible mouse strains. For example, PD-1 or PD-L1 blockade using MoAbs rapidly precipitates diabetes in pre-diabetic female non-obese diabetic (NOD) mice regardless of age, whereas CTLA-4 blockade induces disease only in neonates.[21] Similarly, in the experimental autoimmune encephalitis (EAE) model, PD-1 blockade accelerates and worsens disease, with increased CNS lymphocyte infiltration.[22] Thus, the PD-1 pathway appears to negatively regulate both B cell and T cell immune responses, as well as myeloid cells, and likely plays a role in maintaining peripheral tolerance to autoantigens.

Detailed studies have shown that triggering PD-1 suppresses IL-2 production and the subsequent T cell proliferative response to TCR/CD3-stimulation. Consequently, PD-1 and CTLA-4 both feedback to suppress the T cell proliferative immune response by inhibiting IL-2 production. Extensive studies have been performed to ascertain the mechanism(s) whereby these co-inhibitory receptors mediate their effects. Ligation of both CTLA-4 and PD-1 block CD3/CD28-mediated up-regulation of PKB activity, but each accomplish this regulation using separate, synergistic mechanisms.[23] CTLA-4-mediated inhibition of PKB phosphorylation is sensitive to okadaic acid, providing direct evidence that the serine/threonine phosphatase PP2A plays a prominent role in mediating CTLA-4 suppression of TCR signaling of IL-2 gene expression. Because CD28-mediated activation of Bcl-xL gene expression is not inhibited by CTLA-4 triggering, but is inhibited by PD-1 ligation, CTLA-4 presumably activates PP2A directly, which then dephosphorylates PKB, and the subsequent activation of the IL-2 gene via blocking the activation of the Rel family of transcription factors, especially c-Rel, as well as GSK-3BP and glucose metabolism. By comparison, PD-1 signaling inhibits PKB phosphorylation by preventing CD28-mediated activation of PI3K. Furthermore, the ability to suppress PI3K/PKB activation is dependent upon the immunoreceptor tyrosine-based switch motif located in its cytoplasmic tail, which is not present in the CTLA-4 molecule.

Using DNA microarrays to assess the transcriptional profile after 24 hours of TCR/CD28 triggering, comparing either CTLA-4 or PD-1 suppression, CTLA-4 engagement was found to reduce by ~67% the number of transcripts regulated more than five-fold as a result of TCR/CD28 stimulation, whereas PD-1 triggering reduced ~90% of transcripts.[23] In this regard, the investigators qualified their interpretations of these results by stating that it is important to note that the numbers of CTLA-4 and PD-1 molecules on the T cell surface, as well as the relative affinities these molecules have for their triggering ligands would influence these results. Thus, this is another example of the number of ligand/receptor interactions per unit time that serve as the critical determinants affecting the biological response under investigation.

Early on, when it was discovered that CTLA-4 is activated by the same ligands that activate CD28 (B7.1 and B7.2 molecules expressed by APCs), it was assumed that competition for activating ligands was the primary mechanism whereby CTLA-4 functions to attenuate the T cell immune response. This notion was further enhanced when it was found that the affinity for the B7 ligands is ~10-fold higher for CTLA-4 compared with CD28. Thus, since CD28 is constitutively expressed by T cells, while CTLA-4 is TCR/IL-2-induced, it appears that at least a portion of the negative regulatory effects of CTLA-4 are mediated via competition for the co-stimulatory signals emanating from B7/CD28 signaling.

By comparison, the ligands that trigger PD-1, termed PD-L1 and PD-L2, are distinct from the B7 molecules, even though they belong to the same molecular family based upon their primary structures. These ligands do not trigger either CD28 or CTLA-4. PDL-1 is expressed on resting and up-regulated on activated B, T, myeloid, and DCs, as well as microvascular endothelial cells expressed in non-lymphoid organs such as heart and placenta, in addition to some tumor cells. Thus, it has a much wider distribution than the B7 molecules, which are restricted to APCs.

To delineate the potential functional roles of PD-L1, gene deleted mice were created.[24] PD-L1 (-/-) mice are viable, born at the expected frequency, and appear macroscopically normal with normal microscopic histology of all tissues. The primary and secondary lymphoid tissues are also normal, with normal numbers of thymocytes, CD4+ and CD8+ T cells, B cells, and APCs. Moreover, there was no evidence of spontaneous activation of PD-L1 (-/-) T cells or B cells in mice from 4–16 weeks of age, as assessed using activation surface markers.

Since the T cell also expresses PD-L1 that can trigger a negative feedback signal via its own receptor, to understand the functional significance of PD-L1 expression by T cells, purified CD4+ and CD8+ T cells were studied *in vitro*. When stimulated with solid-phase α-CD3 and soluble α-CD28, the PD-L1 (-/-) T cells produced more IFNγ, especially at low MoAb concentrations, compared with WT T cells. A point of interest with regards to proliferative expansion, there was no difference in IL-2 production between the WT and PD-L1 (-/-) T cells. In a similar fashion, stimulation of WT BALB/c CD4+ T cells with C57Bl/6 PD-L1 (-/-) DCs compared with WT DCs, resulted in the production of greater quantities of IFNγ, but equal amounts of IL-2 and IL-4, as well as proliferation. In addition, there was no difference in proliferation of WT and PD-L1 (-/-) B cells when stimulated with α-IgM, α-CD40, or LPS. However, *in vivo* the antigen-specific expansion of PD-L1 (-/-) CD8+ T cells was doubled compared with WT cells, while antigen-specific CD8+ T cell cytolysis/cell was equal. Accordingly, PD-L1/PD-1 interaction appears to have an important role in limiting the expansion and/or survival of antigen-activated CD8+ T cells, but does not increase the differentiation of the T cell, at least as monitored by cytolytic capacity.

Like CTLA-4, which reacts with two similar but distinct ligands, the PD-1 receptor is stimulated by two similar but distinct ligands that have been conserved despite the evolution of man and mouse, over 75 million years. The PD-L2 molecule was identified via gene bank searching, and found to have 38% amino acid identity to PD-L1, while murine and human PD-L2 have 70% amino acid identity. By comparison with PD-L1, the expression of PD-L2 is more restricted, being confined primarily to IFNγ-, but not TNFα-activated APCs. Again, like PD-L1, PD-L2 functions to inhibit T cell proliferative responses to α-CD3/CD28, as well as to stimulation by specific peptide antigen.

Evidence that the PD-1 ligands are not redundant has come from studies of the PD-L1 (-/-) mice. Although these mice do not develop spontaneous autoimmune disease, mice immunized with myelin oligodendrocyte glycoprotein peptide develop an accelerated and more severe experimental allergic encephalomyelitis (EAE) compared with WT mice. Thus, PD-L2 was not able to compensate for the absence of PD-L1 in this model for the development of this autoimmune disease.

Recently, the tertiary structures of the ligands when bound to their respective receptors in this co-inhibitory family have been determined.[25-28] The PD-1 ectodomain contains a single Ig-like domain typical of the CD28 family, whereas PD-L1 and PD-L2 are composed of two Ig-like domains typical of the B7 family. The structures show a 1:1 ligand/receptor stoichiometry, with interaction primarily between the faces of the Ig domains. In the crystal structures, PD-1 and its ligands are monomers, by comparison with B7 binding to CD28 and CTLA-4, which occurs in dimers. B7-1 is a rigid rod with a very similar structure, whether alone, or bound to CTLA-4. In contrast, the two Ig domains of PD-L1 are in a straight line when complexed with PD-1, but diverge 38° from straight in the absence of PD-1.

How binding of ligand triggers the co-inhibitory receptors to produce a signal on the cytoplasmic side of the membrane remains an unanswered question. Both molecular pairs (i.e. B7/CTLA-4 and PD-L1-2/PD-1) are found in the Immunological Synapse, and should span ~140 Å, which is compatible with the dimensions of the pMHC/TCR ligand/receptor pair. When comparing the uncomplexed structures with the ligand/receptor pairs, there is only a minimal movement of 1–2 Å, thought to be too insufficient to produce a signal. Thus, even though we know the structures of the external domains of these molecules and how they bind to one another, and even though we know that negative signals are generated as a consequence of the intermolecular interaction, we still do not know how the biochemical signals are generated on the cytoplasmic side of the membrane.

Further data that CTLA-4 and PD-1 are separate, thereby regulating T cells distinctly, indicate that PD-1 expression upon T cell activation is regulated by NF-ATc1.[29] Thus, as this TF is activated directly by TCR/CD3 signaling, it appears that the negative feedback loop of this co-inhibitory receptor is activated more directly than the IL-2-mediated expression of CTLA-4.

B and T lymphocyte attenuator (BTLA) is an additional co-inhibitory receptor that has recently been discovered.[30] Because BTLA is also a member of the Ig superfamily, and because it contains two ITIMs in its cytoplasmic domain, which bind tyrosine phosphatases SHP-1 and SHP-2, thereby attenuating IL-2 expression, it was initially assumed that BTLA represented a third member of the CTLA-4/PD-1 co-inhibitory family. However, a ligand found to bind and activate BTLA was found to be a member of the TNF receptor superfamily, which is also known to serve as a herpes virus entry mediator (HVEM).[31] In addition, BTLA is also a co-stimulator, by virtue of its binding to the cell surface protein LIGHT, and as a receptor for lymphotoxin-α (LT-α).[32] Thus, BTLA, HVEM, LIGHT, LT-α and the LIGHT receptor (LTβR) constitute a complex regulatory network. Moreover, another member of the Ig superfamily, CD160, has been found to bind HVEM and deliver a potent inhibitory signal to T cells. Accordingly, the molecular co-inhibition of T cell activation is a complex and rapidly moving field of research, and future developments undoubtedly will impact any interpretations and conclusions drawn at this time.

References

1. Smith, K. (1982) Lymphokines as lymphocytotrophic hormones. In *Lymphokines.* E.A.L. Pick, M, editor. Academic Press. New York, NY. pp. 203–211.
2. Smith, K.A. (1990) Interleukin 2: The first hormone of the immune system. *Scientific American* **262**:50–57.
3. Medawar, P. (1973) *Immunopotentiation.* Assoc. Sci. Pub. Amsterdam, The Netherlands.
4. Robb, R.J., Munck, A., and Smith, K.A. (1981) T cell growth factor receptors: quantitation, specificity, and biological relevance. *J. Exp. Med.* **154**:1455–1474.
5. Smith, K.A., Favata, M.F., and Oroszlan, S. (1983) Production and characterization of monoclonal antibodies to human interleukin 2: strategy and tactics. *J. Immunol.* **131**:1808–1815.
6. Thompson, C.B., Lindsten, T., Ledbetter, J.A., Kunkel, S.L., Young, H.A., Emerson, S.G., Leiden, J.M., and June, C.H. (1989) CD28 activation pathway regulates the production of multiple T-cell-derived lymphokines/cytokines. *Proc. Natl. Acad. Sci. USA* **86**:1333–1337.
7. Fraser, J.D., Irving, B.A., Crabtree, G.R., and Weiss, A. (1991) Regulation of interleukin-2 gene enhancer activity by the T cell accessory molecule CD28. *Science* **251**:313–316.
8. Jenkins, M., and Schwartz, R. (1987) Antigen presentation by chemically modified splenocytes induces antigen-specific T cell unresponsiveness *in vitro* and *in vivo. J. Exp. Med.* **165**:302–319.
9. Quill, H., and Schwartz, R. (1987) Stimulation of normal inducer T cell clones with antigen presented by purified Ia molecules in planar lipid membranes: specific

induction of a long-lived state of nonresponsiveness. *J. Immunol.* **138**:3704–3712.

10. Brunet, J., Denizot, F., Luciani, M., Roux-Dosseto, M., Suzan, M., Mattei, M., and Goldstein, P. (1987) A new member of the immunoglobulin superfamily-CTLA-4. *Nature* **328**:267–270.

11. Walunas, T., Lenschow, D., Bakker, C., Linsley, P., Freeman, G., Green, J., Thompson, C., and Bluestone, J. (1994) CTLA-4 can function as a negative regulator of T cell activation. *Immunity* **1**:405–413.

12. Tivol, E., Borriello, F., Schweitzer, A., Lynch, W., Bluestone, J., and Sharpe, A. (1995) Loss of CTLA-4 leads to massive lymphoproliferation and fatal multiorgan tissue destruction, revealing c critical negative regulatory role of CTLA-4. *Immunity* **3**:541–547.

13. Waterhouse, P., Penningger, J., Timms, E., Wakeman, A., Shahinian, A., Lee, K., Thompson, C., and Mak, T. (1995) Lymphoproliferative disorders with early lethality in mice deficient in CTLA-4. *Science* **270**:985–988.

14. Alegre, M., Noel, P., Eisfelder, B., Chuang, E., Clark, M., Reiner, S., and Thompson, C. (1996) Regulation of surface and intracellular expression of CTLA-4 on mouse T cells. *J. Immunol.* **157**:4762–4770.

15. Ishida, Y., Agata, Y., Shibahara, K., and Honjo, T. (1992) Induced expression of PD-1, a novel member of the immunoglobulin gene superfamily, upon programmed cell death. *EMBO J.* **11**:3887–3895.

16. Agata, Y., Kawasaki, A., Nishimura, H., Ishida, Y., Tsubata, T., Yagita, H., and Honjo, T. (1996) Expression of the PD-1 antigen on the surface of stimulated mouse and B lymphocytes. *Int. Immunol.* **8**:765–772.

17. Carter, L., Fouser, L., Jussif, J., Fitz, L., Deng, B., Wood, C., Collins, M., Honjo, T., Freeman, G., and Carreno, B. (2002) PD-1: PDL inhibitory pathway affects both CD4$^+$ and CD8$^+$ T cells and is overcome by IL-2. *Eur. J. Immunol.* **32**:634–643.

18. Nishimura, H., Minato, N., Nakano, T., and Honjo, T. (1998) Immunological studies on PD-1-deficient mice: implication of PD-1 as a negative regulator for B cell responses. *Int. Immunol.* **10**:1563–1572.

19. Nishimura, H., Nose, M., Hiai, H., Minato, N., and Honjo, T. (1999) Development of lupus-like autoimmune diseases by disruption of the PD-1 gene encoding an ITIM motif-carrying immunoreceptor. *Immunity* **11**:141–151.

20. Nishimura, H., Okazaki, T., Tanaka, Y., Nakatani, K., Hara, M., Matsumori, A., Sasayama, S., Mizoguchi, A., Hiai, H., Minato, N., *et al.* (2001) Autoimmune dilated cardiomyopathy in PD-1 receptor-deficient mice. *Science* **291**:319–322.

21. Ansari, M., Salama, A., Chitnis, T., Smith, R., Yagita, H., Akiba, H., Yamazaki, T., Azuma, M., Iwai, H., khoury, S., *et al.* (2003) The programmed death-1 (PD-1) pathway regulates autoimmune disease in nonobese diabetic (NOD) mice. *J. Exp. Med.* **198**:63–69.

22. Salama, A., Chitnis, T., Imitola, J., Akiba, H., Tushima, F., Azuma, M., Yagita, H., Sayegh, M., and Khoury, S. (2003) Critical role of the programmed death-1 (PD-1) pathway in regulation of experimental autoimmune encephalitis. *J. Exp. Med.* **198**:70–78.

23. Parry, R., Chemnitz, J., Frauwirth, K., Lanfranco, A., Braunstein, I., Kobayashi, S., Linsley, P., Thompson, C., and Riley, J. (2005) CTLA-4 and PD-1 receptors inhibit T cell activation by distinct mechanisms. *Mol. Cell. Biol.* **25**:9543–9553.

24. Latchman, Y., Liang, S., Wu, Y., Chernova, T., Sobel, R., Klemm, M., Kuchroo, V., Freeman, G., and Sharpe, A. (2004) PD-L1-deficient mice show that PD-L1 on T cells, antigen-presenting cells, and host tissues negatively regulates T cells. *Proc. Natl. Acad. Sci. USA* **101**:10691–10696.

25. Stamper, C.C., Zhang, Y., Tobin, J.F., Erbe, D.V., Ikemizu, S., Davis, S.J., Stahl, M.L., Seehra, J., Somers, W.S., and Mosyak, L. (2001) Crystal structure of the B7-1/CTLA-4 complex that inhibits human immune responses. *Nature* **410**:608–611.

26. Lin, D.-W., Tanaka, Y., Iwasaki, M., Gittis, A., Su, H.-P., Mikami, B., Okazaki, T., Honjo, T., Minato, N., and Garboczi, D. (2008) The PD-1/PD-L1 complex resembles the antigen-binding Fv domains of antibodies and T cell receptors. *Proc. Natl. Acad. Sci. USA* **105**:3011–3016.

27. Lazar-Molnar, E., Yan, Q., Cao, E., Ramagopal, U., Nathenson, S., and Almo, S. (2008) Crystal structure of the complex between programmed death-1 (PD-1) and its ligand PD-L2. *Proc. Natl. Acad. Sci. USA* **105**:10483–10488.

28. Ikemizu, S., Gilbert, R., Fennelly, J., Collins, A., Harlos, K., Jones, E., Stuart, D., and Davis, S. (2000) Structure and dimerization of a soluble form of B7-1. *Immunity* **12**:51–60.

29. Oestreich, K.J., Yoon, H., Ahmed, R., and Boss, J.M. (2008) NFATc1 regulates PD-1 expression upon T cell activation. *J. Immunol.* **181**:4832–4839.

30. Watanabe, N., Gavrieli, M., Sedy, J.R., Yang, J., Fallarino, F., Loftin, S.K., Hurchla, M.A., Zimmerman, N., Sim, J., Zang, X., *et al.* (2003) BTLA is a lymphocyte inhibitory receptor with similarities to CTLA-4 and PD-1. *Nat. Immunol.* **4**:670–679.

31. Sedy, J.R., Spear, P.G., and Ware, C.F. (2008) Cross-regulation between herpes viruses and the TNF superfamily members. *Nat. Rev. Immunol.* **8**:861–873.

32. Krieg, C., Boyman, O., Fu, Y.X., and Kaye, J. (2007) B and T lymphocyte attenuator regulates CD8$^+$ T cell-intrinsic homeostasis and memory cell generation. *Nat. Immunol.* **8**:162–171.

Chapter 14

The Paradox of the IL-2 (-/-) Mouse

Given the pivotal role of the IL-2/IL-2R interaction in driving both the proliferative clonal expansion of antigen-selected T cells, as well as the negative feedback regulation of TCR-signaled IL-2 production, it follows that loss of such a high fidelity control of the systemic immune response will inevitably lead to immune system diseases. In particular, mutations in the genes encoding the critical molecules of the negative feedback loops should result in a lowering of the threshold of the discrimination of self versus non-self recognition, so that autoimmune diseases might ensue. Actually, we have already touched on the severe lethal autoimmune syndrome observed when the genes encoding CTLA-4 are deleted.

When considering these issues, it is important to reiterate the Quantal Theory and to underscore that there is a critical number of TCRs that must be triggered before a T cell is activated to express the genes encoding IL-2 and its receptor chains. Furthermore, there is also a critical number of IL-2Rs that must be triggered before a cell makes the irrevocable quantal decision to replicate its DNA and undergo cytokinesis. In this regard, remaining unexplored is the relationship between the critical number of triggered IL-2Rs and negative feedback signaling versus positive proliferative signaling. In other words, we do not understand how the positive and negative signals are integrated.

We have also seen how exquisitely sensitive the tremendous number of clonal TCRs (i.e. mean = 10^4/cell) are to a small number of non-self peptides, even 1 peptide/cell, in the face of thousands of self-peptide molecules presented via MHC molecules on an APC. T cells encounter self-peptides on APCs continuously, but do not react. On the one hand, the density of individual self-peptides presented by an individual APC must be low. Otherwise self-reactivity would be more prevalent than it is. On the other hand, one other major and obvious difference between non-self versus self-peptides is the duration that they can interact with the immune system. By definition self-peptides persist, while non-self-peptides are transient and cleared by the immune system. Therefore, there must be mechanisms that have evolved and have been selected for, which prevent reactivity to persisting self-antigens.

Tachyphylaxis, a phenomenon commonplace in neuroscience and pharmacology, describes a situation whereby continuous signaling via a ligand/receptor pair eventually results in attenuation of signals transmitted. Various mechanisms can be operative, from ligand-mediated accelerated internalization and degradation of the receptor, as we have seen to occur in the case of IL-2 and its receptor, to "desensitization" of the receptor, so that it no longer signals. Desensitization can be due to several different molecular mechanisms, including activation of negative feedback loops that abrogate signaling. We have already discussed the SHP-1 phosphatase negative feedback loop that is operative in TCR signaling, and the CTLA-4/PD-1 negative feedback loops operative to dampen TCR signaling of IL-2 gene expression. All of these mechanisms fall under the rubric of tachyphylaxis. The suppressors of cytokine signaling (SOCS) proteins are another example of molecular mechanisms that serve to turn off signaling via the cytokine receptors.

The earliest example of tachyphylaxis in the immune system was the demonstration of antigen-mediated down-regulation of the B cell antigen receptors (BCR).[1,2] Thus, the progeny of mice transgenic for Ig genes reactive with hen egg-white lysozyme (HEL) mated with mice made transgenic for the expression of HEL, permitted for the first time an experimental model to examine the generation of self-tolerance. Many cells are deleted via a central tolerance mechanism, but some remain and are rendered anergic via a tachyphylaxis manifested by down-regulation of surface IgM but not IgD. In this regard, a critical threshold of antigen concentration is necessary to provoke the tolerant or anergic state, thereby suggesting that some sort of counting mechanism occurs and that receptor occupancy per unit time is important in deciding the fate of these cells.

With regard to T cells and tachyphylaxis, the paradox of the IL-2 knockout mouse became paradigmatic. The IL-2 genes were the first cytokine genes to be deleted, reported in 1991 by Ivan Horak and colleagues.[3] Because IL-2 was known to be essential for T cell proliferation after antigen-activation *in vitro*, it was expected that deletion of the IL-2 genes would result in a severe T cell immunodeficiency *in vivo*. Also, the expression of the IL-2Rα-chain (CD25) on immature, double-negative (i.e. CD4-CD8-) thymocytes had been interpreted as indicating that the IL-2/IL-2R interaction could be important for T cell development in the thymus. However, targeted disruption of the IL-2 genes resulted in IL-2 (-/-) mice with apparently normal lymphocyte development during embryogenesis.[3] At birth there were no apparent abnormalities of the number or types of either T cells or B cells in either the primary or secondary lymphoid organs. As anticipated, when cells from young animals were examined, *in vitro* T cell proliferative responses to mitogens such as Con-A were decreased, on average by about 70% compared with WT responses. However, it was astonishing that T cell proliferative responses were not totally absent based on previous *in vitro* experiments. Thus, the IL-2 (-/-) mouse reopened the question as to whether the TCR itself might be capable of promoting T cell proliferation, and that IL-2 simply functioned as an amplifier of a process that was essentially TCR initiated. Alternatively, perhaps other interleukins could substitute for IL-2, albeit not as effectively.

When challenged *in vivo* with the viral pathogens LCMV and VSV in 1993 by Rolf Zinkernagel and Hans Hengartner in collaboration with Ivan Horak, the IL-2 (-/-) mice were immunocompromised compared with WT or IL-2 (+/-) control mice as expected, with the generation of CTL reactivity decreased by ~2/3 compared to WT mice, but again, surprisingly, B cell and T cell immune responses were not totally absent.[4] Moreover, the IL-2 (-/-) mice recovered from their infections. In these early experiments, the data reported were derived from individual mice. Even so, the investigators stated: "[t]hese normal *in vivo* immune responses question the importance of IL-2 as defined by *in vitro* studies."

These findings were interpreted as indicative of an interleukin redundancy, perhaps attributable to cytokines like IL-4 or interleukins yet to be discovered. Even so, IL-2 still appeared to be the principle T cell growth factor *in vivo*. In 1995 Christine Biron's team used several IL-2 (-/-) mice/group, rather than just one mouse, when comparing them with WT mice, and revealed that the usual marked proliferative expansion of WT CD8+ T cells in response to LCMV infection was completely absent in the IL-2 (-/-) mice. Moreover, the generation of LCMV-specific cytolytic T lymphocyte (CTL) responses in IL-2 (-/-) mice were depressed by 90% compared with CTL from WT control mice.[5]

These findings of both *in vitro* and *in vivo* immunodeficiency of the IL-2 (-/-) mice were entirely consistent with the known function of IL-2 in T cell biology, as a T cell growth factor, even though they did not show a total absence of immunoreactivity. However, when these mice aged beyond six weeks a shocking paradoxical polyclonal lymphoproliferative autoimmune syndrome appeared, manifested by the accumulation of activated T cells in multiple organs, including salivary glands, lungs, kidneys, heart, pancreas and liver. Moreover, autoimmune hemolytic anemia (AHA) and inflammatory bowel disease (IBD) ultimately led to premature death.[6] Thus, although similar to the polyclonal lymphoproliferative syndrome of CTLA-4 (-/-) mice, there were distinct differences, especially the kidney infiltration, and the AHA and IBD suffered by the IL-2 (-/-) mice but not by the CTLA-4 (-/-) mice. Even so, these findings led to the hypothesis that an unanticipated crucial defect resulting from the elimination of IL-2 might be the lack of an IL-2-dependent negative feedback function.[7]

Experiments to try to understand this phenomenon demonstrated that IL-2 administration to IL-2 (-/-) mice prevented the onset of the autoimmune syndrome. IL-2 administered twice daily from day three after birth completely prevented the onset of hemolytic anemia and premature death. Furthermore, adoptive transfer of splenocytes and thymocytes from IL-2-treated IL-2 (-/-) mice delayed the onset of the disease, but did not prevent it. Accordingly, it appeared that IL-2 was capable of inducing some maturational event in T cells that somehow prevented the cells from responding to self-antigens.[8] However, whatever this maturational change entailed, it was transient.

References

1. Goodnow, C.C., Crosbie, J., Adelstein, S., Lavoie, T.B., Smith-Gill, S.J., Brink, R.A., Pritchard-Briscoe, H., Wotherspoon, J.S., Loblay, R.H., Raphael, K., *et al.* (1988) Altered immunoglobulin expression and functional silencing of self-reactive B lymphocytes in transgenic mice. *Nature* **334**:676–682.
2. Goodnow, C.C., Crosbie, J., Jorgensen, H., Brink, R.A., and Basten, A. (1989) Induction of self-tolerance in mature peripheral B lymphocytes. *Nature* **342**:385–391.
3. Schorle, H., Holtschke, T., Hunig, T., Schimpl, A., and Horak, I. (1991) Development and function of T cells in mice rendered interleukin-2 deficient by gene targeting. *Nature* **352**:621–624.
4. Kundig, T.M., Schorle, H., Bachmann, M.F., Hengartner, H., Zinkernagel, R.M., and Horak, I. (1993) Immune responses in interleukin-2-deficient mice. *Science* **262**:1059–1061.
5. Cousens, L.P., Orange, J.S., and Biron, C.A. (1995) Endogenous IL-2 contributes to T cell expansion and IFN-gamma production during lymphocytic choriomeningitis virus infection. *J. Immunol.* **155**:5690–5699.
6. Sadlack, B., Lohler, J., Schorle, H., Klebb, G., Haber, H., Sickel, E., Noelle, R.J., and Horak, I. (1995) Generalized autoimmune disease in interleukin-2-deficient mice is triggered by an uncontrolled activation and proliferation of CD4[+] T cells. *Eur. J. Immunol.* **25**:3053–3059.
7. Horak, I., Lohler, J., Ma, A., and Smith, K. (1995) Interleukin-2 deficient mice: a new model to study autoimmunity and self-tolerance. *Immunol. Rev.* **148**:35–44.
8. Klebb, G., Autenrieth, I.B., Haber, H., Gillert, E., Sadlack, B., Smith, K.A., and Horak, I. (1996) Interleukin-2 is indispensable for development of immunological self-tolerance. *Clinical Immunology and Immunopathology* **81**:282–286.

Chapter 15

The *Scurfy* Mouse

In 1949 a spontaneous mutation occurred in a mouse colony at Oak Ridge Tennessee.[1] The mutation was propagated and reported in 1959 to be X-linked and recessive, leading to a failure to thrive syndrome in males, manifested by ruffled fur and weight loss, hence the designation *scurfy*. More than 30 years later, in the early 1990s, the *scurfy* mouse was demonstrated to be suffering from an autoimmune lymphoproliferative syndrome very similar to that of the IL-2 (-/-) mouse and the CTLA-4 (-/-) mouse.[2,3] Remarkably, the syndrome is very acute with an onset soon after birth and a fully developed disease before maturity, leading to death by three weeks of age. Gross morphologic lesions of the scurfy syndrome include runting, scaly skin, squinted eyes, hepatosplenomegaly, and enlarged lymph nodes. The characteristic histological finding is a lymphohistiocytic proliferation and infiltration that effaces lymph node architecture, thickens the dermis, and forms nodular germinal center-like accumulations in the portal areas of the liver. Like the IL-2 (-/-) mouse, the scurfy males suffer from a severe autoimmune hemolytic anemia, and a marked polyclonal hypergammaglobulinemia. A factor of crucial importance, scurfy heterozygous females (i.e. X^{sf}/X^+) are entirely normal, so that normal T cells appear to somehow "*suppress*" the T cells that carry the mutation from becoming hyperactive, in that the extra X chromosome in females should be randomly inactivated in early embryogenesis. Therefore, 50% of the T cells should contain the mutant gene, while 50% should contain a normal gene.

Then new experiments attributed the scurfy phenotype to an abnormality primarily of CD4[+] T cells.[4-6] Whereas treatment with

MoAbs reactive with CD8[+] T cells neither lessened the severity of lesions nor accelerated the disease onset in scurfy mice, treatment with anti-CD4[+] MoAbs alleviated the severity of lesions and significantly increased lifespan. However, additional experiments revealed that other cell types, in particular CD8[+] T cells, can eventually mediate scurfy disease. Thus, mice bred to be both scurfy and CD4-deficient appear phenotypically normal at weaning (three weeks), whereas mice heterozygous for CD4[+] T cells usually succumb to disease before this time and display pathology typical of scurfy disease. Even so, although fairly normal at three weeks of age, CD4-deficient scurfy mice begin exhibiting signs of scurfy disease at four weeks, including enlarged lymph nodes, a high spleen/body weight ratio, reactive T and B cell areas in lymphoid organs, and elevated serum IgG levels, and die at approximately six weeks of age.[5] Thus, their lifespan is only doubled when CD4[+] T cells are absent. These data are extremely important, given that defective CD4[+] T cells have more recently been implicated as the sole perpetrators of the scurfy disease.

Experiments focused on understanding the pathophysiology of scurfy disease implicated over-expression of cytokine genes.[6] Using Northern blots, quantitative PCR and *in situ* hybridization to examine the lymphoproliferative tissues in diseased scurfy mice, increased expression of IL-2, IL-4, IL-5, IL-6, IL-7, IL-10, IFNγ, and TNFα was observed, thereby providing a molecular basis for the inflammatory pathology for the first time. Additional studies revealed an increased production of GM-CSF and Mac-1[+] cells and a relative decrease in B cells. Examination of CD4[+] T cells from scurfy mice revealed expression of cell surface activation markers, including CD25, CD69, and the co-stimulatory ligands, B7-1 and B7-2.[7] Moreover, like the CTLA-4 (-/-) T cells, scurfy T cells exhibit significant proliferation *in vitro* without TCR stimulation, while stimulation with submitogenic concentrations of anti-CD3 revealed increased proliferation, and a decreased requirement for co-stimulation with anti-CD28. These data were interpreted as suggesting that "the scurfy mutation results in a defect that interferes with the normal down-regulation of T cell activation."[7]

These experiments clearly pointed to an intrinsic defect in T cells with mutant scurfy genes, which was traced to an unrestrained expression of cytokine genes. However, the phenotype is not solely due to unrestrained cytokine gene expression by scurfy T cells, in that as already mentioned, female heterozygous scurfy mice (i.e. X^{sf}/X^+) do not succumb to scurfy disease.[1] In this instance, because of the random inactivation of the X chromosome, ~50% of the T cells should contain the scurfy mutation, while the other 50% of T cells should contain a normal allele. Evidently, the presence of normal T cells somehow prevents or suppresses the T cells containing the mutant cells either from developing or from becoming hyperactivated. This interpretation is also consistent with the observation that $X^{sf}O$ females also succumb to scurfy disease.[3] Moreover, scurfy disease can be transferred to T cell deficient nude mice and SCID mice by T cells or thymic transplants from scurfy mice, while scurfy disease does not occur when euthymic mice are given scurfy T cells or thymic transplants.[4] Thus, pathogenic scurfy T cell activities can be inhibited or prevented in immunocompetent recipient mice. One would think that scurfy T cells, capable of "spontaneous" proliferation, would preferentially expand, whether or not there was a lymphopenic environment, and whether or not there were also normal T cells present.

References

1. Russell, W., Russell, L., and Gower, J. (1959) Exceptional inheritance of a sex-linked gene in the mouse explained on the basis that the X/O sex-chromosome constitution is female. *PNAS* **45**:554–560.
2. Godfrey, V., Wilkinson, J., Rinchik, E., and Russell, L. (1991) Fatal lymphoreticular disease in the scurfy (sf) mouse requires T cells that mature in a sf thymic environment: potential model for thymic education. *PNAS* **88**:5528–5532.
3. Godfrey, V., Wilkinson, J., and Russell, L. (1991) X-linked lymphoreticular disease in the scurfy (sf) mutant mouse. *Am. J. Path.* **138**:1379–1387.
4. Godfrey, V., Rouse, B., and Wilkinson, J. (1994) Transplantation of a T cell-mediated lymphoreticular disease from the scurfy (sf) mouse. *Am. J. Path.* **145**:281–286.
5. Blair, P., Bultman, S., Haas, J., Rouse, B., Wilkinson, J., and Godfrey, V. (1994) CD4+CD8- T cells are the effector cells in disease pathogenesis in the scurfy (sf) mouse. *J. Immunol.* **153**:3764–3774.

6. Kanangat, S., Blair, P., Reddy, R., Deheshia, M., Godfrey, V., Rouse, B., and Wilkinson, E. (1996) Disease in the scurfy (sf) mouse is associated with over-expression of cytokine genes. *E. J. Immunol.* **26**:161–165.
7. Clark, L., Appleby, M., Brunkow, M., Wilkinson, J., Ziegler, S., and Ramsdell, F. (1999) Cellular and molecular characterization of the scurfy mouse mutant. *J. Immunol.* **162**:2554–2564.

Chapter 16

Lymphopenia, Autoimmunity and the Regulatory T Cell (Treg)

Early in the investigation of the role of the thymus in the immune system, in 1962 J.F.A.P. Miller showed that although thymectomy of adult mice leads to no discernible abnormalities, neonatal thymectomy within the first three days of life (d3Tx) results in dramatic pathology. For the first four weeks of life, d3Tx mice appear grossly normal and grow identically with sham-thymectomized control mice. However, thereafter the d3Tx mice fail to thrive and lose weight, becoming runted with the gross appearance of the IL-2 (-/-) mice, CTLA-4 (-/-) mice and scurfy mice, and die prematurely from a syndrome characterized by wasting and diarrhea.[1] It is also noteworthy that neonatal thymectomy results in mice that are lymphopenic, even though they appear grossly normal for the first few weeks of life. Moreover, they are immunocompromised, and are unable to mount an antibody response to immunization with *Salmonella typhi* or to reject allogeneic or even xenogeneic skin grafts. Accordingly, these d3Tx mice are definitely immunocompromised, so it was paradoxical that they should die from a wasting disease that resembles graft versus host disease (GvHd), with infiltration of many organs with activated lymphocytes, and Miller emphasized this point, in particular.

A possible explanation for these confusing observations was introduced by Sakaguchi and co-workers, who reported 20 years later, in 1982, that d2-4Tx mice as well as athymic nude (*nu/nu*) mice develop spontaneous tissue-specific autoimmune disorders, such as oophoritis, gastritis, thyroiditis and orchitis when reconstituted with T cells lacking the Lyt-1[+] subset.[2,3] They hypothesized that removing

131

the thymus early in life might result in deprivation of a critical subset of T cells that normally might function to *"suppress or regulate"* potential self-reactive clones. They went on to show that nude mice or d2-4Tx mice reconstituted with Lyt-1$^+$ T cells (which at the time were thought to mark a helper T cell subset, but are not synonymous with CD4$^+$ T cells) could prevent the development of the autoimmune disorders. They went on to speculate that the depletion of this *"regulatory T cell"* subset by d3Tx might result in the post-thymectomy autoimmune diseases.

Subsequently, more than a decade later, Sakaguchi and co-workers reported in 1995 that T cells expressing the α-chain of the IL-2R (CD25) can prevent a lethal lymphoproliferative syndrome caused by CD25$^-$ T cells when transferred to lymphopenic athymic nude (*nu/nu*) mice.[4] Experiments reconstituting nude mice with CD4$^+$ T cells from normal mice, which were depleted of CD25$^+$ T cells revealed that all recipients spontaneously develop histological and serological evidence of a polyclonal multi-organ autoimmune disease, including thyroiditis, gastritis, insulinitis, sialoadenitis, adrenalitis, oophoritis, glomerulonephritis, and polyarthritis. However, if CD4$^+$CD25$^+$ T cells are co-transferred with the CD25$^-$ cells, the autoimmune disease was prevented. In this regard, it is noteworthy that Sakaguchi found that *CD8$^+$ T cells that expressed C25 were also effective in preventing the autoimmune syndrome*, although they were reported to be less effective on a per cell basis than CD4$^+$CD25$^+$ T cells. This observation will take on greater importance as the transcriptional regulator FOXP3 becomes synonymous with Regulatory T cells.

Sakaguchi's group followed up this report with similar experiments using d3Tx mice as recipients of the cells.[5] They first established that CD25$^+$ T cells are not detectable at birth, while adult mice have ~10% CD4$^+$CD25$^+$ T cells and <1% CD8$^+$CD25$^+$ T cells. Instead, CD25$^+$ T cells only appear in the secondary lymphoid tissues after day three in the neonatal period, with a rapid accumulation to near adult levels by two weeks after birth. They then showed that d3Tx retards the accumulation of both CD3$^+$ T cells as well as CD3$^+$CD25$^+$ T cells in the spleen by an order of magnitude, so that like Miller 30 years before them, they confirmed that d3Tx mice are T cell lymphopenic, and especially that they are CD25$^+$ T cell deficient.

Sakaguchi's group then went on to show that the autoimmune diseases that occur spontaneously in d3Tx mice can be prevented by the administration of graded numbers of splenic T cell populations given one week after thymectomy, when monitored for signs of disease after three months. When monitoring histological and serological findings of gastritis, total CD4$^+$ T cells and CD4$^+$CD25$^-$ T cells did not prevent gastritis, but CD4$^+$CD25$^+$ clearly did. However, it is noteworthy that CD8$^+$CD25$^+$ T cells were also effective, although on a per cell basis, less so. Also notable is the fact that equal numbers of CD4$^+$CD25$^+$ T cells and CD8$^+$CD25$^+$ T cells were not tested for their efficacy in this model, so that the authors were careful not to designate whether the effective CD25$^+$ T cells were from the CD4$^+$ or CD8$^+$ T cell subsets.

From these results, it was hypothesized that the peripheral T cell population of d3Tx mice might be deficient in "autoimmune-preventative CD25$^+$ T cells," but still may contain some self-reactive T cells, and mice older than three days of age may bear pathogenic self-reactive T cells, but have insufficient numbers of CD25$^+$ autoimmune preventative T cells to control the self-reactive T cells. To test this hypothesis, T cell populations from BALB/c *nu/+* of various ages were inoculated into nude recipients, which were examined for autoimmune disease three months later. T cells from three-day-old mice could not prevent autoimmunity (i.e. gastritis, oophritis, thyroiditis), while T cells from two-week-old, eight-week-old, or one-year-old mice were effective in preventing disease. However, they were not effective if first depleted of CD25$^+$ T cells using MoAb + C'. Also, using RT-PCR, three-day CD4$^+$ T cells did not express CD25, nor IL-4, IL-10, or TGFβ, and were similar to adult CD4$^+$CD25$^-$ T cells. By comparison, adult CD4$^+$CD25$^+$ T cells expressed transcripts for CD25, IL-2, IL-4, IL-10, and IFNγ, thereby suggesting that these cells are pre-activated *in vivo*.

These data were interpreted as supporting the hypothesis that "one aspect of self-tolerance is maintained by CD25$^+$ T cells that sustain potentially pathogenic self-reactive T cells in an (*inactivated*), CD25⁻, dormant state." Moreover, the "ontogenic time course of these CD25⁺ T cells is 'intrinsically programmed,' as abnormalities in the developmental process (e.g. as in d3Tx), can cause autoimmune disease in "genetically susceptible individuals." Accordingly, these two papers from Sakaguchi's group established the phenomenon that came to be known as regulatory T cells (Tregs), and CD4⁺CD25⁺ markers became accepted as their phenotype. However, it is noteworthy that although Sakaguchi emphasized CD4⁺CD25⁺ T cells in his reports, CD8⁺CD25⁺ T cells also seemed to have regulatory capacities. Also, as already emphasized, CD25 is the IL-2R α-chain, and is a marker of antigen-activated, IL-2-stimulated T cells, so that it could not be used as a strict identification maker for Tregs. Finally, it is noteworthy that Sakaguchi pursued the notion that a "Regulatory" T cell subset existed for over 14 years, changing the identification of the subset, as new T cell surface markers became available. During this interval, the experimental system remained the same, i.e. adoptively transferring T cells to lymphopenic mice.

Accordingly, the commonality in all of these autoimmune disease models is lymphopenia. Thus, lymphopenia, which leads to immunodeficiency, somehow eventually results in a *paradoxical* lymphoproliferation and loss of self-discrimination. In this regard, it is important to note that newborn mice are relatively lymphopenic, especially in peripheral lymphoid organs.[6] During the first few weeks of life, when the animal undergoes rapid growth until puberty at ~5 weeks of age, the thymus is very actively seeding the periphery with mature CD4⁺ and CD8⁺ naive T cells that have undergone "positive selection" as a consequence of TCR/self-pMHC interaction in the thymus. As mentioned above, during the first two weeks of life, the proportion of CD4⁺CD25⁺ cells gradually increases in the periphery to plateau at the adult levels of ~10% of CD4⁺ T cells. Thus, the idea has come to be accepted that if this maturational process is prevented by d3Tx, the residual T cells that are already present in the periphery at birth, proliferate and ultimately react against self-antigens leading to multiorgan immunopathology, *because* there are inadequate numbers of CD4⁺CD25⁺ Tregs to suppress them.

Throughout the pre-pubertal period, the cellularity of the spleen and lymph nodes increases in size as the body grows. Detailed studies have shown that TCR/self-pMHC interaction is required to maintain the thymic-derived naive T cells in the periphery, as is IL-7, in that the cells die if placed in MHC (-/-) or IL-7 (-/-) hosts.[7] Studies have also been performed with other lymphopenic hosts, such as RAG (-/-), SCID or nude mice. If small numbers of normal naive T cells are transferred to these mice, the donor T cells proliferate slowly, switch to a memory phenotype (i.e. CD44[hi]), and gradually the total peripheral T cell pool eventually achieves a near-normal size. Moreover, the stimuli operative in these T-depleted hosts promoting the slow proliferation appear to be the same as those responsible for maintaining the T cell survival and slow proliferation in young mice, as well as normal adult mice, i.e. TCR/self-pMHC and IL-7. Reducing T cell numbers to low levels results in elevated concentrations of IL-7 because it is no longer metabolized and degraded at the normal rate. IL-7-driven homeostatic proliferation is typically slow, so that most cells divide only one to five times over the course of two weeks, and it is more pronounced for CD8[+] T cells than for CD4[+] T cells. However, it is important to note that T-depleted mice reconstituted with normal T cell populations that contain both CD25[-] and CD25[+] T cells do not develop lymphoproliferative autoimmune diseases.

References

1. Miller, J. (1962) Effect of neonatal thymectomy on the immunological responsiveness of the mouse. *Proc. Roy. Soc. London B* **156**:415–428.
2. Sakaguchi, S., Takahashi, T., and Nishizuka, Y. (1982) Study on cellular events in post-thymectomy autoimmune oophoritis in mice. I. Requirement of Lyt-1 effector cells for oocytes damage after adoptive transfer. *J. Exp. Med.* **156**:1565–1576.
3. Sakaguchi, S., Takahashi, T., and Nishizuka, Y. (1982) Study on cellular events in post-thymectomy autoimmune oophoritis in mice. II. Requirement of Lyt-1 cells in normal female mice for prevention of oophoritis. *J. Exp. Med.* **156**:1577–1586.
4. Sakaguchi, S., Sakaguchi, N., Asano, M., Itoh, M., and Toda, M. (1995) Immunologic self-tolerance maintained by activated T cells expressing IL-2 receptor alpha-chains (CD25). Breakdown of a single mechanism of self-tolerance causes various autoimmune diseases. *J. Immunol.* **155**:1151–1164.

136   *The Quantal Theory of Immunity*

5. Asano, M., Toda, M., Sakaguchi, N., and Sakaguchi, S. (1996) Autoimmune disease as a consequence of developmental abnormality of a T cell subpopulation. *J. Exp. Med.* **184**:387–396.
6. Min, B., McHugh, R., Sempowski, G., Mackall, C., Foucras, G., and Paul, W. (2003) Neonates support lymphopenia-induced proliferation. *Immunity* **18**:131–140.
7. Sprent, J., Cho, J.-H., Boyman, O., and Surh, C. (2008) T cell homeostasis. *Immunology and Cell Biology* **86**:312–319.

Chapter 17

Treg-mediated "Active Suppression" of T Cell Proliferation

Because the IL-2Rα chain (CD25) is a specific marker for antigen-activated effector T cells, it was not immediately obvious how and why CD4⁺CD25⁺ T cells could prevent the activation of potential self-pMHC-reactive T cells in lymphopenic d3Tx mice. The internal environment in these lymphopenic animals, which have elevated concentrations of IL-7, is clearly abnormal, and the self-pMHC/IL-7-driven expansion of self-reactive clones could very well contribute to the pathological attack on normal tissues observed in these mice. In this regard, it is worthy of mention that the only known regulators of IL-2Rα chain (CD25) expression are the TCR/CD3/CD28 activation of the Rel TFs and the IL-2-dependent activation of STAT5, with STAT5 playing a much larger role than the Rel TFs. It is also noteworthy that the other γ_c cytokines, IL-4, IL-7, IL-9, IL-15, and IL-21 do not compensate for the loss of IL-2 signals, even though they also all activate STAT5. Thus, the IL-2 (-/-), IL-2Rα (-/-) and IL-2Rβ (-/-) mice all develop a similar immunocompromised/lymphoproliferative autoimmune syndrome.[1-3] By comparison, deletion of the other γ_c cytokines does not lead to similar autoimmune abnormalities. Accordingly, there is something extraordinary about IL-2 signaling that is not shared by the other γ_c cytokines, even though they may activate similar signaling pathways.

By comparison with lymphopenic mice where there is a complete absence of T cells (i.e. RAG (-/-), SCID and nude), the IL-2 (-/-), IL-2Rα (-/-), and IL-2Rβ (-/-) mice, all have peripheral T cells, at least initially. Over time, as T cells accumulate in these animals, the concentrations of T cell-derived cytokines increase. Moreover, these animals also have elevated concentrations of the stromal-derived cytokines, such as IL-7 and IL-15. Thus, the IL-2 (-/-) mice have elevated concentrations of IL-4 and IL-21, in addition to elevated concentrations of IL-7 and IL-15. Moreover, the IL-2Rα (-/-) and IL-2Rβ (-/-) mice have elevated concentrations of IL-2, because it is not being internalized and degraded normally (for review, Ref. 4). In the IL-2Rα (-/-) mice, the compensatory increase in IL-2 concentrations can drive the proliferative expansion of T cells activated by self-pMHC complexes via interaction with the IL-2Rβ/γ heterodimer, which contains the cytoplasmic domains responsible for signaling, and which has an intermediate affinity for IL-2 binding ($K_d = 1$ nM), as well as IL-15 produced by supporting stromal cells.

These issues become very important when considered in the light of a report from Shevach's group, which essentially solidified the notion that CD4+CD25+ T cells have regulatory activity, and function to prevent autoreactivity by T cells that escape negative selection in the thymus.[5] As enunciated by Shevach, two fundamentally different models have been proposed to explain the association between lymphopenia and autoimmunity. In both models, normal adult mice contain autoreactive T cells.

1) "In the '*empty space*' model, the paucity of T cells in the peripheral lymphoid organs permits the expansion of the precursors of autoreactive T cells, because the T lymphopenic environment facilitates the interaction of autoreactive T cells with professional APCs. Such (*self-antigen*)-activated autoreactive T cells could then migrate to non-lymphoid organs and induce organ-specific autoimmunity."

It is noteworthy that there is no mention by Shevach of cytokines as contributors of the abnormal "lymphopenic environment." Evidently, at that time (1998) he ascribed to the notion that antigens are solely responsible for mediating the proliferation of antigen-selected T cells.

2) In the second model, the *"Regulatory T cell subset deficiency"* model, which Shevach attributes to Sakaguchi (6), "the lymphopenic state results in the *selective absence* of a critical population of regulatory T cells that normally continuously suppress the activation of autoreactive T cells."

Because of the requirement that one must cause autoimmunity by transferring a population of disease-inducing T cells into a lymphopenic environment, Shevach felt that such an environment might also contribute to the pathogenesis of the Treg deficiency model. As Shevach's 1998 publications confirmed and extended Sakaguchi's findings, it is important to consider them in detail.

Shevach's group repeated Sakaguchi's experiments, transferring CD25$^-$ T cells into both nude and d3Tx mice, monitoring the animals for autoimmune gastritis by histopathology scoring of inflammatory infiltrates into the gastric mucosa. He was able to induce gastritis and interpreted his results as indicative of a defect in the d3Tx mice that could not be secondary solely to the "empty space" generated by the removal of the thymus. Moreover, he was able to prevent the autoimmune gastritis by injecting 2–3 × 10^7 normal spleen/LN cells on day 10 after thymectomy, but not by the same cell population depleted of CD25$^+$ T cells using MoAb + C'. However, Shevach was concerned that it was not clear whether the CD4$^+$CD25$^+$ T cells were derived *in vivo* in response to normal antigenic activation, or whether the CD4$^+$CD25$^+$ T cells were a unique population of *"professional immunoregulatory cells."* As one means to test this question, TCR Tg T cells on a SCID mouse background were used as potential suppressors. Since the antigenic specificity of these TCR Tg T cells was against chicken ovalbumin, which presumably could not cross-react with gastric autoantigens, these cells should not "regulate" the autoimmune reaction, which was found to be the case.

However, since the Tg T cells contained ~four-fold fewer CD25$^+$ T cells, this quantitative difference could have led to their incapacity to suppress the gastritis. Consequently, the d3Tx mice were reconstituted with TCR Tg and then immunized with chicken ovalbumin +

complete Freund's adjuvant (CFA). Although this maneuver resulted in expression of CD25 on ~50% of the Tg T cells, no inhibition of gastritis was observed as monitored by histopathology scoring. Thus, Shevach argued that these results supported the concept of a "professional regulatory cell" present in the peripheral lymphoid tissues of normal mice. Moreover, the data indicated that the expression of CD25 on a large proportion of cells was insufficient to suppress the autoimmune reaction, thereby supporting the notion that the regulation by Tregs could not be simply ascribed to the passive consumption and metabolism of IL-2 by the IL-2R$^+$ cells.

These data were interpreted as indicating that the Treg cells could suppress the activation of autoreactive T cells. To address the hypothesis that Tregs could also suppress the effector function of already activated autoreactive cells, T cell clones reactive to gastric parietal cell H/K ATPase were generated from d3Tx mice. Transfer of these cloned cells to nude mice, but not normal BALB/c mice, induced gastritis. It was not explored as to why normal mice were not susceptible to gastritis, but to test whether this autoreactivity could be suppressed by Tregs, 5×10^6 autoreactive cloned T cells were transferred to nude recipients together with $50–100 \times 10^6$ normal BALB/c splenocytes (i.e. 10–20-fold excess). This maneuver completely prevented gastritis in the nude recipients. Shevach interpreted these data as consistent with the capacity of Tregs to also suppress the functions of activated effector T cells. However, he could not prove that CD25$^+$ cells in the normal splenocyte populations were responsible for the suppressive effect, in that it was "difficult to purify enough CD25$^+$ cells to treat more than a few animals."[5] Even so, Shevach concluded that the CD4$^+$CD25$^+$ subpopulation could suppress both the induction and the effector functions of autoreactive T cells.

This report was complimented by another report from the Shevach group that focused on examining the characteristics of Tregs *in vitro*.[7] Using an *in vitro* culture system of CD4$^+$CD25$^-$ T cells from both BALB/c and C57Bl/6 mice and T cell-depleted irradiated splenocytes (therefore B cells, macrophages and DCs) as accessory cells, activated with suboptimal concentrations of anti-CD3 (0.5 μg/mL), it was established that the CD4$^+$CD25$^+$ peripheral subset in normal adult mice is a potent inhibitor of *polyclonal* T cell proliferation. Furthermore, they showed that the suppression is mediated by a cytokine-independent, but cell-contact-dependent mechanism that requires activation of the $CD4^+CD25^+$ T cell via the TCR. Moreover, they showed that the mechanism of suppression involved inhibition of IL-2 mRNA expression by the responding $CD4^+CD25^-$ effector cells, so that the suppression can be circumvented by supplementing the cultures with exogenous IL-2 or by enhanced IL-2 production by co-stimulation with anti-CD28 MoAbs. The claim that the suppression is dependent upon cell-cell contact is based on experiments whereby the T_{eff} cells and the Treg cells are separated via a transwell apparatus.

Because there was not, and still is not, a surface marker capable of distinguishing $CD4^+CD25^+$ Tregs from TCR/CD3-activated $CD4^+CD25^+$ effector T cells, it is important to examine this report in detail, since the definition of a Treg cell subsequently became based on this *in vitro* polyclonal functional suppressor assay. Like Sakaguchi's findings, ~10% of peripheral $CD4^+$ T cells were found by Shevach's group to be $CD25^+$, and were mixed in their expression of other "activation" markers. Thus, ~50% were $CD69^+$ and $CD45RB^{lo}$, and most were $CD62L^{hi}$, indicative of an activated phenotype. Of interest, even though these purified $CD4^+CD25^+$ cells expressed the IL-Rα chain, they were unresponsive even to "high" IL-2 concentrations (3 ng/mL = 200 pM, a concentration that is just saturating for high affinity trimeric IL-2Rs). In addition, purified $CD4^+CD25^+$ T cells were *anergic* in that they did not proliferate in response to stimulation with soluble or solid phase anti-CD3, anti-CD3/28, or Con-A, even when accessory cells were present, while purified $CD4^+CD25^-$ T cells proliferated in response to these stimuli.

Since IL-2, but not the TCR or CD3/28, mediates anti-CD3-induced proliferation, these data suggested that the anergy of $CD25^+$ T cells is due to the incapacity to produce IL-2. Thus, if both anti-CD3 and IL-2 are supplied, then both $CD25^+$ and $CD25^-$ T cell subsets proliferate equally well, which confirms that the anergy is secondary to the incapacity to produce IL-2, but not the incapacity to respond to IL-2. This conclusion was further supported by data showing that if the early signaling events of TCR/CD3 signaling are by-passed via activation with phorbol myristic acetate (PMA) + ionomycin, or by PMA + IL-2, CD4+CD25+ cells can proliferate.

Given that CD4+CD25+ T cells cannot proliferate in response to either soluble or solid-phase anti-CD3, while CD4+CD25− cells can, it became possible to test whether the CD25+ cells could suppress the proliferation of CD25− cells. As shown in Fig. 17.1, CD4+CD25− T cells cultured at 50,000 cells/well and stimulated by soluble anti-CD3 (1.0 µg/mL), proliferated and incorporated 75,000 CPM of ^{3}H-TdR after three days of culture. However, as few as 5,000 CD4+CD25+ T cells (i.e. a 1:10 ratio of CD25+:CD25− T cells) decreased the ^{3}H-TdR by ~75%, and 25,000 CD25+ T cells (a 1:2 ratio) suppressed the proliferation of the CD25− T cells completely. However, when stimulation was performed via solid-phase anti-CD3, which provides an optimal TCR/CD3 activation signal, the CD25+ T cells

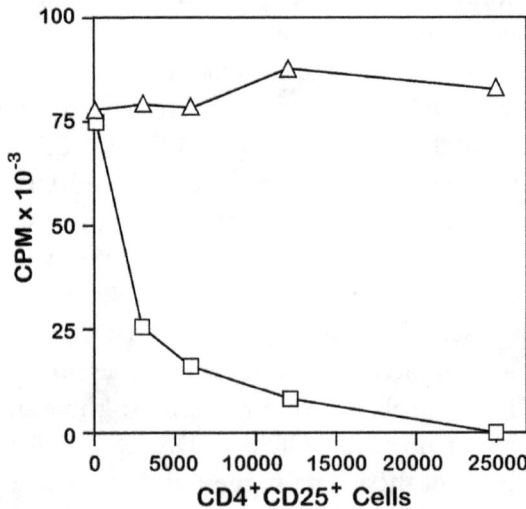

Figure 17.1:   CD4+CD25+ T cells suppress the proliferation of CD4+CD25− T cells. CD4+CD25− T cells (5 × 10⁴) were incubated with plate-bound anti-CD3 (△), or with 1.0 µg/mL soluble anti-CD3 (□) in the presence of accessory cells (T-depleted splenocytes, 5 × 10⁴), and the indicated number of CD4+CD25+ T cells. (Redrawn from: Thornton, A.M. and Shevach, E. 1998. *J. Exp. Med.* **188**:287–296.)

were not suppressive at all, even at 25,000 cells/well. Also, an important point is that the CD25$^+$ cells were not suppressive unless activated via the TCR as shown by experiments with antigen-specific CD25$^-$ T cells and CD25$^+$ cells. As TCR stimulation promotes the expression of the trimeric IL-2R, as well as markedly increasing the expression of the IL-2R α-chain, these data are consistent with the possibility that in order to be suppressive, the Treg cells must be incapable of producing IL-2, but capable of binding, internalizing and catabolizing IL-2, which is mediated only by the trimeric IL-2R, but not by CD25 alone.[8]

Experiments focused on possible molecular mechanisms of Treg "active suppression" of effector T cells revealed that inhibitory cytokines could not be responsible. It is worthy of mention in this regard, that Sakaguchi's group found that even unstimulated CD4$^+$CD25$^+$ T cells express mRNA transcripts for the proliferation-inducing cytokines, IL-2, IL-4, but also the inhibitory cytokines, TGFβ, and IL-10.[9] In contrast, Shevach's group found that unstimulated CD4$^+$CD25$^+$ T cells did not express detectable mRNA transcripts for IL-2 and IL-4, but did express IL-10 and TNFα transcripts. Moreover, even after anti-CD3 stimulation, CD4$^+$CD25$^+$ T cells did not express detectable IL-2 mRNA transcripts, a result that Shevach states "is consistent with the failure of CD25$^+$ T cells to proliferate when stimulated with anti-CD3," whereas anti-CD3-stimulated CD25$^-$ T cells expressed readily detectable IL-2 transcripts, and did proliferate.

One other aspect of the 1998 Thornton–Shevach report that subsequently became very important for the definition of Treg cells, relates to their cell contact-dependent mechanisms of action. The CD25$^-$ T cells (5×10^5) were cultured in 24-well plates with *maximal* concentrations of soluble anti-CD3 (3.0 μg/mL) and accessory cells (T-depleted splenocytes, 5×10^5). Then, increasing concentrations of CD25$^+$ T cells were added, either mixed together with the CD25$^-$ cells, or separated by a cell-impermeable membrane via a transwell apparatus, in the absence or presence of accessory cells. The most effective suppression occurred when the Tregs and T effectors were cultured together in the presence of accessory cells, although examination of the one experiment of this kind presented, showed that there was suppression to ~50% even when the Tregs were separated from the CD25⁻ cells. Even so, these findings were interpreted to indicate that the mechanism of suppression entailed cell-cell contact, although which cells had to make contact was left open, because there were always three cell types included, CD25$^+$ and CD25$^-$ T cells, and APCs.

Since CD25$^+$ T cells have been reported to suppress IL-2 production by CD25$^-$ T cells via cell-cell contact, the data in support of this contention are very important. To examine this suppressive effect of CD25$^+$ T cells, CD25$^-$ T cells (5×10^4) were cultured with equal numbers of accessory cells (5×10^4) activated by *submaximal* concentrations of soluble anti-CD3 (0.5 μg/mL) in the presence or absence of CD25$^+$ T cells (2.5×10^4; 1:2 suppressors:effectors). Supernatants were collected after 24, 48, and 72 hours, to monitor for secreted IL-2 by ELISA, as shown in Fig. 17.2A. It is important to remember that IL-2 production is transient after anti-CD3 activation, with maximal concentrations appearing by 24–48 hours, followed by its rapid disappearance thereafter, which is shown in this experiment, when CD4$^+$ T cells alone are activated. The maximal IL-2 concentration produced in this experiment, ~3 ng/mL = 200 pM, is just saturating for the high affinity trimeric IL-2R. When CD25$^-$ T cells are co-cultured at a 2:1 ratio with CD25$^+$ T cells, a 10-fold lower concentration of IL-2 is detectable after 24 hours of culture (~20 pM), which is at the EC$_{50}$ for cells with high affinity trimeric IL-2Rs. Moreover, IL-2 was undetectable after 48 hours. However, this experiment cannot exclude the possibility that the CD25$^+$ T cells might consume the IL-2 produced by the CD25$^-$ T cells, especially within the first 24 hours, before the IL-2R expression by the T$_{eff}$ becomes maximal.

Therefore, to circumvent this criticism, RNA was purified from CD4$^+$ T cells stimulated in the presence or absence of CD25$^+$ T cells after 16 hours of stimulation with soluble anti-CD3 (Fig. 17.2B). In this experiment, IL-2 mRNA could not be detected, although in other experiments, "a slight band could be detected." In this regard, it is noteworthy that IL-2 mRNA expression is transient in these circumstances, with a peak at ~6 hours, and it usually is undetectable

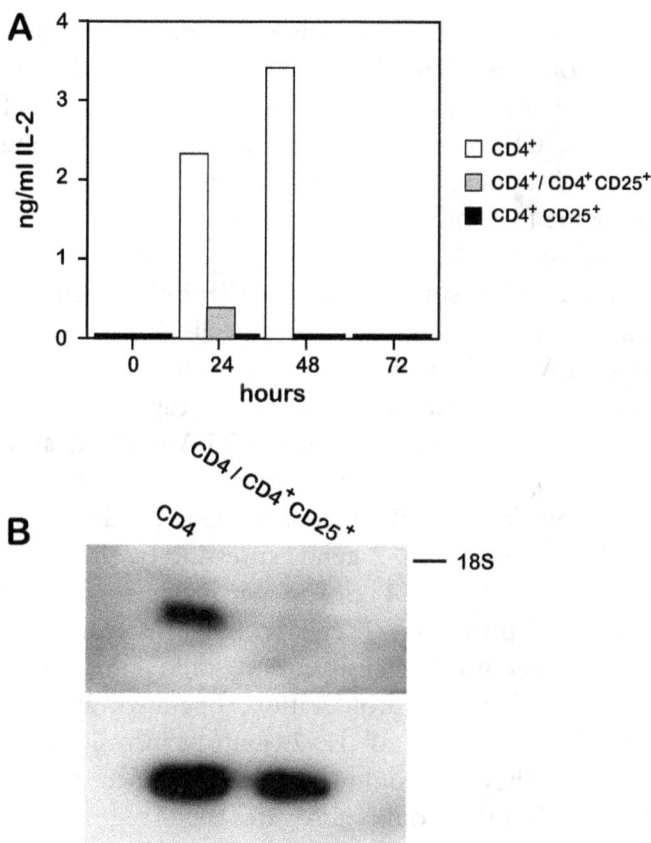

Figure 17.2: CD4+CD25+ T cells suppress IL-2 production by CD4+CD25− T cells.
A. CD4+CD25− T cells (5 × 10⁴), accessory cells (5 × 10⁴), and 0.5 µg/mL soluble anti-CD3 were cultured in the presence or absence of CD4+CD25+ T cells (2.5 × 10⁴) (2:1 25−/25+). Supernatants were taken at the indicated times and IL-2 quantified by ELISA (LLD 5 pM). **B.** RNA was purified from CD4+ T cells stimulated in the presence or absence of CD4+CD25+ T cells after 16 hours. Top, IL-2 mRNA; bottom, β-actin. (Redrawn from: Thornton, A.M. and Shevach, E. 1998. *J. Exp. Med.* **188**:287–296.)

after 12 hours. Moreover, if the CD4+CD25+ cannot express the IL-2 gene, to compare the mRNA expression with and without the presence of the CD4+CD25+ T cells would have the effect of diluting the IL-2 mRNA expressed by the CD4+CD25− T cells. However, on the basis of this one experiment, it was concluded that "these results strongly support the view that the CD25⁺ population exerts its inhibitory effects by blocking the TCR/CD3-induction of IL-2 production by the CD25⁻ cells." Most noteworthy, according to the authors, "these findings indicated that the Treg-mediated block occurs *before* the level of IL-2 transcription, *and was dependent upon the presence of the ACs.*"

One of the other surface markers that subsequently came to be identified with CD4⁺CD25⁺ Treg cells is the co-inhibitory molecule CTLA-4. Thus, CTLA-4 expression by the Treg cells themselves could be responsible for the incapacity of these cells to express the IL-2 gene. Moreover, CTLA-4 could also be responsible for suppressing the co-stimulation of CD25⁻ T cells by competitively blocking the B7 ligands on APCs, which are necessary for maximal activation of IL-2 gene expression by the CD25⁻ effector T cells. Since CTLA-4 clearly participates in the negative feedback of IL-2-promoted T cell proliferative clonal expansion, this mechanism is entirely plausible, and could explain the abrogation of polyclonal T cell proliferation, operative at the level of TCR/CD3/28 signaling of IL-2 gene expression. However, Thornton and Shevach could find no evidence that blocking CTLA-4 with MoAbs could abrogate the Treg suppression of proliferation stimulated by anti-CD3.

Thus, four papers established and defined Tregs for the first time. The two Sakaguchi papers established CD4⁺CD25⁺ T cells as a distinct developmental lineage that occurred via thymic maturation within the immediate two weeks of the postnatal period, and that functioned *in vivo* in lymphopenic models of autoimmunity to prevent auto-reactive T cells from invading organs to cause tissue destruction.[9,10] The two Shevach papers then defined these CD4⁺CD25⁺ Treg cells as capable of suppressing the *in vitro* proliferation of effector T cells activated *polyclonally*, by inhibiting IL-2 gene expression via some mechanism that required cell-cell contact and the presence of accessory APCs.[5,7] By this definition, Tregs had to be activated via their TCR to become suppressive, and they were capable of actively suppressing any activated effector T cell, regardless of antigen specificity. At this time, the Tregs were envisioned as acting only to prevent reactivity against auto-antigens, and suppression of reactivity to foreign antigens was not considered. Thus, how an immune response to foreign microbial antigens could occur with Tregs suppressing all antigen reactivity remained unclear.

Also, obviously problematic, IL-2R$^+$ T cells can passively bind, internalize and degrade IL-2, thereby leading to the possibility that the suppression of proliferation could be more apparent than real, because effector T cell proliferation is driven by IL-2. Moreover, problematic with the designation of CD4$^+$CD25$^+$ T cells as Tregs is that the same phenotype is shared by antigen-activated, non-anergic, and non-suppressive effector T cells. In addition, although Sakaguchi had shown that CD8$^+$ cells could also suppress autoimmunity *in vivo*, Shevach presented no data comparing the suppressive activity of CD8$^+$CD25$^+$ T cells with CD4$^+$CD25$^+$ T cells *in vitro*.

"Nevertheless, Tregs and CD4$^+$CD25$^+$ T cells became synonymous in the minds and publications of most immunologists. Thus, the renaissance of the 1970s CD8$^+$ 'suppressor T cells' came forth in a new guise and a new name 30 years later. The lust for suppressor T cells then grasped and overwhelmed a new generation of immunologists, who were armed with a vast supply of cellular and molecular reagents, as well as the power of genetic manipulation. The hope to boost Treg activity to suppress autoimmune/inflammatory diseases, allergies, and allograft rejection, or to block their suppressive governorship of the immune response to augment the immune reactivity for the treatment of cancer, infectious diseases, and as an adjuvant for vaccines, became irresistible."[11]

In this regard, this overwhelming desire for some means to manipulate the immune response, and the hope that the immune system may have evolved mechanisms to do so, seemed so logical. As detailed earlier, negative feedback signaling via CTLA-4 and PD-1 had already been identified as molecules that functioned in this manner. Whether a distinct cell such as a Treg actually exists and the molecular mechanism(s) it utilizes to actively suppress all other cells became the next important questions.

References

1. Sadlack, B., Lohler, J., Schorle, H., Klebb, G., Haber, H., Sickel, E., Noelle, R.J., and Horak, I. (1995) Generalized autoimmune disease in interleukin-2-deficient mice is triggered by an uncontrolled activation and proliferation of CD4+ T cells. *Eur. J. Immunol.* **25**:3053–3059.

2. Suzuki, H., Kundig, T.M., Furlonger, C., Wakeham, A., Timms, E., Matsuyama, T., Schmits, R., Simard, J.J., Ohashi, P.S., Griesser, H., *et al.* (1995) Deregulated T cell activation and autoimmunity in mice lacking interleukin-2 receptor beta. *Science* **268**:1472–1476.

3. Willerford, D., Chen, J., Ferry, J., Davidson, L., Ma, A., and Alt, F. (1995) Interleukin-2 receptor alpha chain regulates the size and content of the peripheral lymphoid compartment. *Immunity* **3**:521–530.

4. Boyman, O., Letourneau, S., Frieg, C., and Sprent, J. (2009) Homeostatic proliferation and survival of naive and memory T cells. *Eur. J. Immunol.* **39**:2088–2094.

5. Suri-Payer, E., Amar, A., Thornton, A., and Shevach, E.M. (1998) CD4+CD25+ T cells inhibit both the induction and effector function of autoreactive T cells and represent a unique lineage of immunoregulatory T cells. *J. Immunol.* **160**:1212–1218.

6. Sakaguchi, S., Fukuma, K., Kuribayashi, K., and Masuda, T. (1985) Organ-specific autoimmune diseases induced in mice by elimination of T cell subset. I. Evidence for the active participation of T cells in natural self-tolerance; deficit of a T cell subset as a possible cause of autoimmune disease. *J. Exp. Med.* **161**:72–87.

7. Thornton, A.M., and Shevach, E.M. (1998) CD4+CD25+ immunoregulatory T cells suppress polyclonal T cell activation *in vitro* by inhibiting interleukin 2 production. *J. Exp. Med.* **188**:287–296.

8. Hemar, A., Subtil, A., Lieb, M., Morelon, E., Hellio, R., and Dautry-Varsat, A. (1995) Endocytosis of interleukin-2 receptors in human T lymphocytes: distinct intracellular localization and fate of the receptor alpha, beta, and gamma chains. *J. Cell Biol.* **129**:55–64.

9. Asano, M., Toda, M., Sakaguchi, N., and Sakaguchi, S. (1996) Autoimmune disease as a consequence of developmental abnormality of a T cell subpopulation. *J. Exp. Med.* **184**:387–396.

10. Sakaguchi, S., Sakaguchi, N., Asano, M., Itoh, M., and Toda, M. (1995) Immunologic self-tolerance maintained by activated T cells expressing IL-2 receptor alpha-chains (CD25). Breakdown of a single mechanism of self-tolerance causes various autoimmune diseases. *J. Immunol.* **155**:1151–1164.

11. Smith, K., and Popmihajlov, Z. (2008) The quantal theory of immunity and the interleukin-2-dependent negative feedback regulation of the immune response. *Immunol. Rev.* **224**:124–140.

Chapter 18

FOXP3, A Better ID-Tag for Tregs?

In 2000, the gene responsible for the scurfy syndrome (*sf*) was cloned by Fred Ramsdell's group in Seattle, Washington in collaboration with J. Wilkinson of the group from Oak Ridge Tennessee, which had originally identified the mutation.[1] The protein encoded by this gene was found to be a new member of the forkhead/winged-helix family of transcriptional regulators, which has over 80 members, and was designated FOXP3. Also, it was found to be highly homologous to an orthologous gene in humans. In scurfy mice, a frame-shift mutation results in a product lacking the carboxy-terminal forkhead domain. Genetic complementation demonstrated that the protein product of FOXP3 could restore normal immune homeostasis.

Coincident with this report, the human FOXP3 gene was found to be mutated in individuals suffering from the X-linked auto-immunity-allergic dysregulation (XLAAD) syndrome,[2] as well as the X-Linked neonatal diabetes mellitus, Enteropathy and PolyEndocrin-ology (IPEX) syndrome.[3,4] Obviously the finding of the same gene mutated in these two syndromes showed them to be identical. Moreover, together with the murine data, these findings solidified the importance of the FOXP3 gene product for normal immunological homeostasis. Individuals born with mutations in this gene suffer from severe autoimmunity, with rapid death similar to the scurfy mouse. Like CTLA-4, PD-1, and IL-2, mutations in the FOXP3 gene markedly upset the capacity of the immune system to remain quiescent in the face of auto-antigens. The mechanistic relationship of these genes then became the question then to ask.

Initial experiments focused on discerning the function of FOXP3 revealed that the forkhead domain is required for nuclear localization and DNA binding.[5] Furthermore, upon over-expression in CD4[+] T cells, FOXP3 attenuates TCR-activation of IL-2 production by T cells, and inhibits the transcription of NF-AT response elements from the IL-2 promoter. Thus, early on it appeared that FOXP3 functions as a transcriptional repressor of IL-2 and perhaps other cytokine genes. Obviously, the mutational crippling of this negative regulatory function could reset the threshold of TCR activation to a lower level, such that normal auto-antigens might be able to trigger activation of cytokine gene expression, thereby initiating self-reactivity. Accordingly, FOXP3 constitutes another molecule that participates in a negative feedback loop, regulating TCR activation of cytokine gene expression. However, unlike CTLA-4 and PD-1, which interfere with TCR/CD3/CD28 signaling at the level of the membrane, FOXP3 interferes at the level of IL-2 transcription.

Because of the similarity of the autoimmune syndromes that result from FOXP3 deficiency and the syndromes that occur in lymphopenic mice reconstituted with CD4[+]CD25[−] T cells, those interested in Tregs examined the expression of FOXP3 by different thymocyte and peripheral T cell subsets, and three reports appeared almost simultaneously in early 2003.[6-8] It was almost too good to be true that Sakaguchi's group found FOXP3 expression to be restricted *primarily* to CD4[+]CD25[+] thymocytes.[6] Moreover, real-time quantitative PCR analyses revealed that FOXP3 mRNA levels were 100-fold higher in the CD4[+]CD25[+] subset of mature peripheral T cells, compared with peripheral CD4[+]CD25[−] T cells. It is noteworthy that at this time, there were no antibodies available to test the single cell expression of FOXP3 protein, so that analysis of mRNA expression by purified T cell subsets was the only means available to examine FOXP3 expression. Moreover, given the fact that CD25 expression is an indicator of recent antigen activation, it is noteworthy that FOXP3 mRNA expression was not influenced by activation via solid-phase anti-CD3, supplemented by either anti-CD28 or IL-2, when assayed daily for three days. Accordingly, Sakaguchi's group concluded that "FOXP3 expression is stable in CD4$^+$CD25$^+$ T cells, irrespective of the mode or state of activation."

To test whether FOXP3 expression would confer a "Regulatory" phenotype to naive T cells, CD4$^+$CD25$^-$ T cells were infected with bicistronic retroviral vectors expressing both FOXP3 and GFP as a marker for infected cells. After one week of culture, activation of these cells with anti-CD3 showed very little proliferation compared with control, non-infected cells and those infected with only GFP-containing vectors, thereby displaying the anergic phenotype of Tregs. Furthermore, FOXP3 expressing cells produced low levels of IL-2, IFNγ, IL-4 and IL-10 by comparison with controls. In this regard, the concentration of IL-2 produced by the control cells was low, only 15 pM, just enough to half-saturate the high affinity IL-2R, while these cells produced ~120 pM IL-4, which is a high concentration for "normal" T cells. In addition, the lack of IL-10 production by the FOXP3$^+$ cells is at variance with the phenotype of Tregs, which Sakaguchi and Shevach had shown previously to be characteristic for Tregs.

Also, when tested for their phenotype, the FOXP3-infected cells rang true for Treg cells. They expressed "higher" levels of CTLA-4, GITR, CD25, and CD103 than the controls. Moreover, when tested for their *in vitro* suppressive activity, only the FOXP3-infected cells suppressed the proliferation of anti-CD3-activated CD25$^-$CD4$^+$ T cells cultured at a 1:1 ratio for 72 hours with APCs. Finally, these FOXP3$^+$-infected cells were capable of suppressing autoimmune gastritis and inflammatory bowel disease *in vivo*.

Rudensky's group essentially confirmed these data using real-time PCR to quantitatively detect FOXP3 mRNA, and western blot analysis with a polyclonal rabbit anti-serum raised against recombinant FOXP3 protein.[7] They found ~40-fold higher amounts of FOXP3 mRNA in CD4$^+$CD25$^+$ T cells by comparison with CD4$^+$CD25$^-$ T cells from C57Bl/6 mice. Like Sakaguchi's group, they also found that activation with solid-phase anti-TCR + anti-CD28 + rIL-2 (6.7 μM, which is 67,000-fold in excess of a high affinity IL-2R saturating IL-2 concentration, i.e. 100 pM) failed to promote FOXP3 mRNA expression by either CD4$^+$CD25$^+$ or CD25$^-$ T cell subsets.

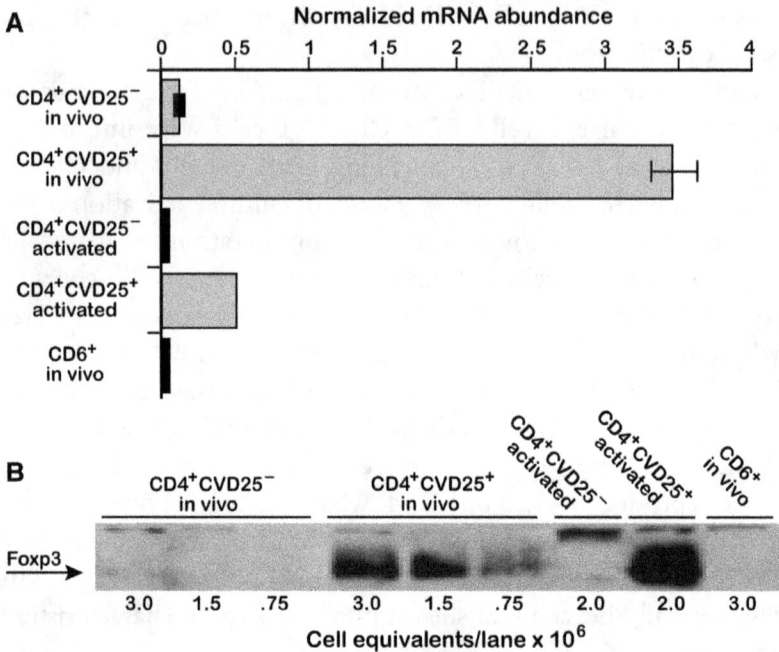

Figure 18.1:   FOXP3 is specifically expressed in CD4⁺CD25⁺ regulatory T cells. A.
Real-time quantitative PCR for FOXP3 mRNA from purified T cell subsets. **B.** Western
blot analysis in purified T cell subsets using rabbit anti-FOXP3 IgG. (Redrawn from:
Fontenot, J. *et al.* 2003. *Nature Immunol.* **4**:330–336.)

Freshly isolated CD8⁺ T cells also failed to express detectable FOXP3
mRNA, but no data were reported for activated CD8⁺ T cells. The
western blot analysis essentially mirrored the PCR data, except that
activated CD4⁺CD25⁺ T cells expressed greater amounts of FOXP3
protein. Again, activated CD8⁺ T cell data were conspicuously absent,
as shown in Fig. 18.1.

By comparison with Sakaguchi's group, which examined the role
of FOXP3 by over-expressing the gene, Rudensky's group used
homologous recombination to generate FOXP3 conditional KOs.
Like the scurfy mutant mice, these mice also succumbed to generalized autoimmunity and died by four weeks of age. The CD4⁺CD25⁺
thymocytes from 28-day-old FOXP3-mice proliferated *in vitro* when

activated with ConA (but not spontaneously), while the same subset from WT mice did not. Moreover, these cells did not suppress the proliferation of CD4+CD25- T_{eff} cells *in vitro*. These data were interpreted as supporting a role for FOXP3 in the development of Tregs.

To explore the role of FOXP3 in Treg development further, mixed BM chimeras were made. Six weeks after BM reconstitution no CD25+FOXP-thymocytes or peripheral LN T cells were detectable, whereas CD25+FOXP3+ cells were detectable in normal numbers, thereby indicating a developmental/survival *disadvantage* to CD4+ T cells that lack FOXP3. These results were interpreted as "providing clear and direct evidence for a requirement of FOXP3 in Treg development." In this regard, it is noteworthy that the FOXP3- T cells did not survive and expand, in that this result may well explain why female X^{sf}/X^+ mice do not develop scurfy disease, despite random X chromosome inactivation, if the T cells that lack FOXP3 do not survive. Thus, the question is whether FOXP3- T cells have an intrinsic defect that prevents their survival and expansion, or whether these defective cells are recognized by normal FOXP3+ Treg, which suppress the FOXP3- cells extrinsically.

In the same issue as the Rudensky group report, a report from Fred Ramsdell's group further established the connection between FOXP3 and Tregs.[8] They noted that in mutant (X^{sf}/Y) mice, which lack a functional FOXP3 gene, almost 40% of CD4+ T cells express CD25, compared with only 10–15% of WT CD4+ T cells. Moreover, all of these CD25+ T cells appeared to be activated, in that they were large cells, and also expressed activation markers such as CD69 and CD62L. Consistent with the hypothesis that only FOXP3+ T cells are suppressive, when tested for suppression of proliferation of WT CD4+ T cells, the addition of sf CD4+CD25+ cells actually resulted in *increased* ^{3}H-TdR incorporation.

To further explore the relationship between FOXP3 expression and suppressor activity, FOXP3 transgenic mice were developed. It was entirely unexpected that twice the proportion of FOXP3 transgenic CD4+ T cells expressed CD25 (~20%) compared with WT CD4+ T cells (~10%). Also, it is especially noteworthy that FOXP3+ CD4+CD25+ T cells, CD4+CD25- T cells, as well as CD8+ T cells, all

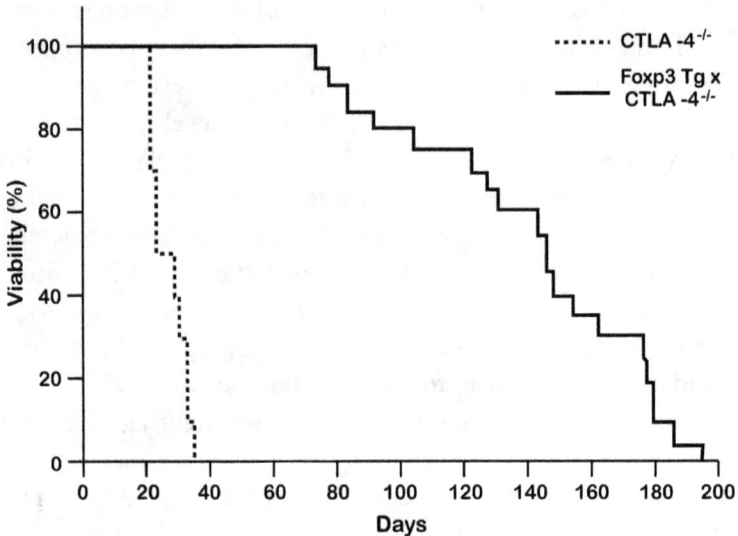

Figure 18.2:   Over-expression of FOXP3 delays lethality in CTLA-4 (-/-) mice.
CTLA-4 (-/-) mice (n = 10) and CTLA-4 (-/-)/FOXP3 Tg mice (n = 20) were observed for signs of disease and mortality over the indicated months. Results are presented as the % viability, which was calculated as 100 × (number of surviving mice/total number of mice) on any given day. (Redrawn from: Khattri, R. *et al.* 2003. *Nature Immunol.* **4**:337–342.)

had suppressive activity. In contrast, FOXP3⁺ B cells were not suppressive, an important exception.

Perhaps the most important experiments from the Ramsdell group were those obtained by crossing CTLA-4 (-/-) mice with FOXP3 Tg mice. As shown in Fig. 18.2, the expression of FOXP3 by CTLA-4 (-/-) mice markedly prolonged their survival, from 50% at ~21 days to 150 days. Moreover, when tested *in vitro*, the suppressive activity of CTLA-4 (-/-) FOXP3⁺ T cells was "generally comparable to that of CD4⁺CD25⁺ T cells from transgenic mice." Thus, it was concluded that CTLA-4 is not required for the generation or suppressive activity of Treg cells. These findings are important in light of more recent publications from Sakaguchi's group, as we shall see.

Accordingly, these three reports solidified and established FOXP3 as the *bona fide* marker for Tregs, and CD4+CD25+FOXP3+ cells became rapidly accepted as the cells imbued with the remarkable capacity to recognize and suppress the polyclonal proliferation of TCR-activated T_{eff} cells.

References

1. Brunkow, M., Jeffrey, E., Hjerrild, K., Paeper, B., Clark, L., Yasayko, S.-A., Wilkinson, J., Galas, D., Ziegler, S., and Ramsdell, F. (2001) Disruption of a new forkhead/winked-helix protein, scurfin, results in the fatal lymphoproliferative disorder of the scurfy mouse. *Nature Genetics* **27**:68–73.
2. Chatlia, T., Blaeser, F., Ho, N., Lederman, H., Voularopoulos, C., Helms, C., and Bowcock, A. (2001) JM2, encoding a forkhead-related protein, is mutated in X-linked autoimmunity-allergic disregulation syndrome. *J. Clin. Invest.* **106**:R75–R81.
3. Wildin, R., Ramsdell, F., Peake, J., Faravelli, F., Casanova, J.-L., Buist, N., Levy-Lahad, E., Mazzella, M., Goulet, O., Perroni, L., *et al.* (2001) X-linked neonatal diabetes mellitus, enteropathy and endocrinopathy syndrome is the human equivalent of mouse scurfy. *Nature Genetics* **27**:18–20.
4. Bennet, C., Christie, J., Ramsdell, M., Brunkow, M., Ferguson, P., Whitesell, L., Kelly, T., Saulsbury, F., Chance, P., and Ochs, H. (2001) The immune dysregulation, polyendocrinopathy, enteropathy, X-linked syndrome (IPEX) is caused by mutations of FOXP3. *Nature Genetics* **27**:20–21.
5. Schubert, L., Jeffrey, E., Zhang, Y., Ramsdell, F., and Ziegler, S. (2001) Scurfin (FOXP3) acts as a repressor of transcription and regulates T cell activation. *J. Biol. Chem.* **276**:37672–37679.
6. Hori, S., Nomura, T., and Sakaguchi, S. (2003) Control of regulatory T cell development by the transcription factor FOXP3. *Science* **299**:1057–1061.
7. Fontenot, J., Gavin, M., and Rudensky, A. (2003) FOXP3 programs the development and function of CD4+CD25+ regulatory T cells. *Nature Immunol.* **4**:330–336.
8. Khattri, R., Cox, T., Yasayko, S.-A., and Ramsdell, F. (2003) An essential role for scurfin in CD4+CD25+ T regulatory cells. *Nature Immunol.* **4**:337–342.

Chapter 19

Mice Versus Men

Confounding this new definition of Tregs were data that accumulated subsequently, concerning FOXP3 expression by human T cells. In contrast to the reports that found FOXP3 expression restricted to murine CD4+CD25+ cells, and not influenced by activation via TCR/CD28 stimulation, peripheral mature human CD4+CD25− T cells activated via TCR/CD28 were found to express FOXP3.[1] Moreover, human CD8+ T cells were also found to express FOXP3.[2,3] Therefore, as FOXP3 mutants of both mice and man results in a very similar fatal autoimmune syndrome, these results called into question the concept that only CD4+CD25+FOXP3+ T cells function to actively suppress potentially self-reactive T cells. Thus, would one need to extend the definition of Tregs to both CD4+ and CD8+ CD25+FOXP3+ T cells? Also, if FOXP3 expression can be activated by TCR/CD28 signaling, does that mean that FOXP3 expression does not delineate a separate cell lineage? Can all T cells become potential Tregs, if they can express FOXP3 and CD25, as well as the other cell surface markers that delineate this phenotype (i.e. CTLA-4, GITR)? Moreover, what are the critical stimuli that lead to FOXP3 expression? In this regard, it is especially noteworthy that early on investigators speculated that the expression of FOXP3 by an antigen-activated T cell could act as a natural feedback loop that would prevent unrestricted cytokine production and inflammatory reactions, particularly in response to auto-antigens.[2]

These early studies on FOXP3 expression were hindered because FOXP3-reactive antibodies were not available, so that the experimenters were relegated to monitoring mRNA expression by *in vitro* separated T cell subsets. Once antibodies became available, Rudensky's group reported that single cell analysis of FOXP3 expression revealed that only ~2–3% of freshly isolated human peripheral blood CD4$^+$ T cells express detectable FOXP3.[4] Moreover, this fraction of cells represented ~10% of the CD4$^+$CD25$^+$ T cells, which in the human was found to be ~20–25% of CD4$^+$ T cells. By comparison, <0.5% of CD8$^+$ T cells expressed FOXP3, and all of these cells were found to be CD25$^-$.

According to Rudensky's group, upon activation of human PBMCs via anti-CD3, a gradual increase in the proportion of FOXP3$^+$ cells occurs, with a peak of ~25% FOXP3$^+$CD4$^+$ and CD8$^+$ T cells after three days of culture. Thereafter there was a gradual decline in the proportion of FOXP3$^+$ cells, so that this increased FOXP3 expression was only transient. Moreover, this "activated" or "induced" FOXP3 expression did not confer the capability of suppressing either IL-2 or IFNγ expression by T$_{eff}$ cells. Thus, these findings further complicated the notion that Tregs are a differentiated lineage, and that FOXP3 is the *sine qua non* of a Treg cell. These findings were subsequently confirmed and extended to show that almost all CD4$^+$ human T cells can be made to express FOXP3, provided they are activated with solid-phase anti-CD3 $^+$ soluble anti-CD28, and both IL-2 and transforming growth factor beta (TGFβ).

In immunology, many things can be observed, especially when dealing with human cells, but concepts are never really accepted as "true" unless and until they are observed in the mouse! Immunologists who only work with murine systems perpetuate this attitude. To a certain extent this attitude is justified, because often definitive experiments are difficult to perform when dealing only with humans or human cells. However, in this instance, since a mutational deficiency of FOXP3 leads to the same lethal autoimmune syndrome in both mice and man, it behooves one to come up with a unifying hypothesis to explain the findings in both species.

Thus, it is particularly noteworthy that subsequent experiments by Ethan Shevach's group found that murine T cells behave similarly. Mature, peripheral CD4$^+$CD25$^-$ T cells activated via CD3/CD28 and supplemented with both IL-2 and TGFβ for several days of culture results in *all* of the cells eventually becoming FOXP3$^+$.[5] Moreover, in contrast to the human "induced" Tregs (iTreg), those generated *in vitro* in the mouse were found to suppress anti-CD3 stimulation of T$_{eff}$ cell proliferation. It is also noteworthy that both TGFβ and IL-2 are required to generate murine FOXP3$^+$ iTregs *in vitro*.

These results led to the notion that there are two types of Tregs. First, those Tregs that arise "naturally" in the thymus, so-called natural Tregs (nTreg), have become thought of as life-long *"policemen of the periphery"*, functioning to preserve peripheral tolerance to self-molecules, and thereby prevent autoimmunity. In addition, a second type of Treg, the antigen-induced iTreg, has been postulated to arise only transiently upon introduction of foreign antigen, to depend upon signals derived from the TCR/CD28, IL-2 and TGFβ, and perhaps to function to dampen an antigen-specific response, whether a foreign or self-peptide. Obviously, the problem inherent in naming all CD4$^+$CD25$^+$FOXP3$^+$ cells Tregs ultimately rests with the two defining functional characteristics of Treg cells, i.e. anergy and "active suppression" of T$_{eff}$ cell proliferation. Moreover, do Treg cells only regulate potential T$_{eff}$ cell responses to self-peptides, or do they function to also suppress T$_{eff}$ cell responses to foreign peptides? Moreover, if Tregs can distinguish between self versus non-self-peptide stimulators, what possibly could be the molecular mechanism(s) of this remarkable ability?

References

1. Walker, M., Kasprowicz, D., Gersuk, V., Benard, A., Landeghen, J., Buckner, J., and Ziegler, S. (2003) Induction of FOXP3 and acquisition of T regulatory activity by stimulated human CD4$^+$CD25$^-$ T cells. *J. Clin. Invest.* **112**:1437–1443.
2. Morgan, M., van Bilsen, J., Bakker, A., Heemskerk, B., Schilham, M., Hartgers, F., Elferink, B., van der Zanden, L., de Vries, R., Huizinga, T., *et al.* (2005) Expression of FOXP3 mRNA is not confined to CD4$^+$CD25$^+$ T regulatory cells in humans. *Hum. Immunol.* **66**:13–20.
3. Wang, J., Ioan-Facsinay, A., van der Voort, E., Huizinga, T., and Toes, R. (2007) Transient expression of FOXP3 in human activated nonregulatory T cells. *Eur. J. Immunol.* **37**:129–138.

4. Gavin, M., Torgerson, T., Houston, E., DeRoos, P., Ho, H., Stray-Pedersen, A., Ocheltree, E., Greenberg, P., Ochs, H., and Rudensky, A. (2006) Single-cell analysis of normal and FOXP3-mutant human T cells: FOXP3 expression without regulatory T cell development. *PNAS* **103**:6659–6664.

5. Davidson, T., Dipaolo, R., Anderson, J., and Shevach, E. (2007) IL-2 is essential for TGF-beta-mediated induction of FOXP3$^+$ T regulatory cells. *J. Immunol.* **178**:4022–4026.

Chapter 20

Active Versus Passive Suppression and IL-2 Metabolism

In Chap. 8, the metabolism of IL-2 and IL-2Rs was reviewed. To recapitulate, only high affinity trimeric IL-2Rs are internalized, and essentially only when triggered by IL-2 binding. The $t_{1/2}$ for internalization is only 15 minutes. The quaternary IL-2/IL-2R complex takes a clatherin-independent internalization pathway, whereby the IL-2R α-chain is recycled to the cell surface, while IL-2 along with the IL-2R β-chains and γ-chains are trafficked to lysosomes and degraded. The end result of this IL-2/IL-2R metabolism is a very rapid diminution of IL-2 concentrations, such that if IL-2 production is discontinued for any reason, IL-2R$^+$ cells rapidly consume and degrade IL-2, thereby providing a built-in mechanism that guarantees cessation of IL-2 signaling of proliferative expansion. Moreover, *in vitro* the cell surface density of the IL-2R α-chains remains high, while the cell surface density of the β-chains and γ-chains decreases to a new steady state of ~50% of the initial densities within two hours.

The dynamics of IL-2 binding, internalization, and metabolism become important if it is accepted that the tempo, magnitude and duration of T cell clonal expansion are IL-2/IL-2R-dependent and not solely determined by pMHC activation of the TCR. Based upon *in vitro* assays, and especially the employment of a transwell system whereby Tregs are separated from the T_{eff} cells, the dogma evolved that Treg cells *actively* suppress T_{eff} cell proliferation by a mechanism that requires cell-cell contact.[1,2] In addition, Shevach's group traced the mechanism to an inhibition of early IL-2 gene expression. As detailed in Chap. 17, in an experiment performed using a mixture of Tregs, T_{eff} cells at a ratio of 1:2, a Northern blot of mRNA extracted after 16 hours of stimulation with a suboptimal anti-CD3 concentration (0.5 μg/mL) found no detectable IL-2 mRNA, while IL-2 mRNA was readily detectable in the T_{eff} cells when activated without Tregs[1] (see Fig. 23, Chap. 17).

Subsequent studies by Shevach's group found that the *in vitro* suppressive activity of Tregs was markedly augmented by IL-2 or IL-4 but not by IL-6, IL-7, IL-9, IL-10, or IL-15.[3,4] Moreover, blocking CTLA-4 or CD28 did not prevent the IL-2/4-augmentation of suppressor function. In follow-up of these reports, Scheffold's group examined the role of IL-2 in the activation of Treg suppressive activity using the pre-activation system of Shevach.[5] They confirmed the original observations of the Shevach group, and then improved upon the assay conditions by using human $CD4^+CD25^-$ responder cells and pre-activated mouse $CD4^+CD25^+$ cells as Tregs. In this way they could selectively block the murine IL-2Rs on the Tregs without influencing the IL-2Rs of the human responder cell population. Blocking MoAbs reactive with murine IL-2Rα/β completely abrogated the suppressive activity, a result that the authors interpreted as: "[t]his clearly demonstrates that the uptake of IL-2 by the Tregs is absolutely required during the suppression." They then went on to show that during the co-culture of Tregs and T_{eff} cells IL-2 is produced in low amounts by the T_{eff} cells and is selectively bound by Tregs. Accordingly, these investigators interpreted their results as consistent with the notion that Tregs need to "take up" IL-2 in order to be suppressive, but they still ascribed to the interpretation that Tregs actively suppress T_{eff} cell IL-2 production, and they further conjectured that IL-2 functioned to enhance the suppressive capacity of Tregs by promoting their production of the inhibitory cytokine IL-10.[6] Alternatively, since murine T cells can bind, internalize and degrade human IL-2, these experimental results are entirely consistent with the "active" suppression of the human T_{eff} cells as actually due to the "passive" consumption of the human IL-2 by the murine IL-2R$^+$ Tregs.

The passive consumption of IL-2 by FOXP3[+] T cells that cannot produce IL-2, but by virtue of a high density of IL-2Rs, can effectively bind, internalize and metabolize IL-2, is one simple molecular mechanism that could explain Treg suppression of T_{eff} cell proliferation. Accordingly, Pushpa Pandiyan working in Michael Lenardo's group examined this hypothesis in detail.[7] To examine Treg cell function, this group used the Shevach assay most commonly used to assess Treg suppressor activity *in vitro*, with either ^{3}[H]-TdR incorporation or dilution of dividing carboxyfluorescein succinimidyl ester (CFSE)-labeled CD4[+]CD25[-]FOXP3[-] responder cells as the read-out. They found that the incorporation of ^{3}[H]-TdR after 36 and 66 hours of culture by the responder cells was suppressed by the presence of Treg cells, as was the dilution of CFSE dye when monitored after 72 hours. Noteworthy was the observation that suppression occurred whether the cells were activated by anti-CD3 + APCs, or by anti-CD3 + anti-CD28, without APCs.

Since decreased ^{3}[H]-TdR incorporation and CSFE dilution could occur because of cell death, rather than simply suppression of proliferation, cell death was assessed by trypan blue staining, as shown in Fig. 20.1. It is readily apparent that the presence of Treg cells resulted in considerable cell death. Comparable results were obtained by flow cytometry with Annexin V and propidium iodide staining, gaiting on the CFSE-labeled responder cells. Also, transmission electron microscopy revealed the classical hallmarks of apoptosis, including condensed nuclei, and membrane "blebbing" of the responder cells. Moreover, "pan-caspace" inhibitors abrogated the cell death. Even so, because measurements of ^{3}[H]-TdR incorporation and CSFE dilution do not clearly distinguish failure of activation and/or proliferation from failure to survive, the Treg and T_{eff} cells were cultured separately. Apoptosis only occurred when the cells were recombined. Moreover, the apoptosis slowly reached a peak three to four days after activation.

Accordingly, all of these data indicated that the Treg cells somehow killed the T_{eff} cells. Experiments to examine whether this was an active or a passive process effectively excluded an active Treg-mediated cytolytic mechanism. Thus, Treg cells from perforin-deficient or Fas

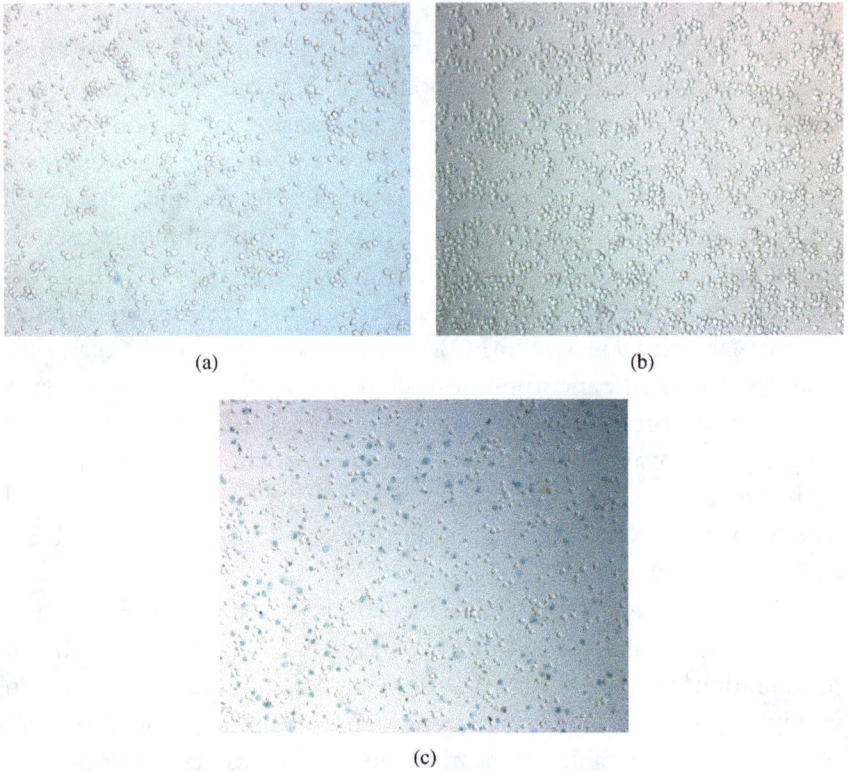

(a)

(b)

(c)

Figure 20.1: Treg cells induce apoptosis of responder T cells *in vitro*. Trypan Blue staining of co-cultures of responder T cells stimulated for three days with anti-CD3 + anti-CD28, alone or with control cells or Treg cells. Original magnification x 200. (From: Pandiyan, P. *et al.* 2007. *Nature Immunol.* **8**:1352–1363.)

ligand mutant mice were just as effective "killers" as cells from WT mice, and inhibitors of TNFα or granzymes had no effect.

As reported previously, the exogenous supplementation of the cultures with IL-2 abrogated the Treg cytolytic effect. Moreover, other IL-2Rγ_c-using cytokines, such as IL-4, IL-7, IL-15, and IL-21 also abrogated the cytolysis, albeit less efficiently than IL-2, thereby suggesting that the apoptosis could be due to cytokine deprivation. Moreover, one of the hallmarks of cytokine deprivation apoptosis, the pro-apoptotic molecule Bad was found activated in responder T cells when Treg cells were present but not in their absence.

These findings were thus consistent with cytokine deprivation apoptosis. When measured directly, IL-2 concentrations were much lower in supernatants of co-cultures of Tregs and T_{eff} cells than in co-cultures of control cells + T_{eff} cells stimulated with anti-CD3 + anti-CD28, even when supplemented with IL-7. Even so, because the lack of IL-2 accumulation could be due either to Treg-mediated suppression of T_{eff} cell IL-2 production or due to IL-2 consumption by the Tregs, steady state *il2* mRNA expression was measured using quantitative PCR in RNA extracted from T_{eff} cells that had been separated from co-cultures of T_{eff} cells + Treg cells. The separation of the T_{eff} from the Tregs before extraction of the RNA was an important step not performed by Thornton and Shevach.[1] Since Tregs cannot produce IL-2, their RNA could have diluted the *il2* mRNA from the T_{eff} cells, thereby producing a false-negative signal, particularly when total RNA, rather than polyA-selected RNA, was monitored by Northern blot analysis. Thus, when measured by the very sensitive quantitative PCR, Pandiyan and co-workers found no difference in responder T_{eff} cell *il2* mRNA whether or not Tregs were present in the co-culture, even when monitored early and frequently over 48 hours of culture. Moreover, in other experiments, *il2* transcription was unaltered by the presence of Tregs, as was the accumulation of intracellular IL-2 protein and the secretion of IL-2 as analyzed by a cytokine capture assay. Accordingly, contrary to the earlier Thornton and Shevach report, Treg cells do not suppress *il2* mRNA expression, nor does there seem to be any effect of Tregs on *il2* transcription, translation or secretion.

Consequently, to account for the lack of IL-2 accumulation in the co-cultures that contained the Treg cells, the most logical conclusion is that Treg cell IL-2 consumption is responsible. One of the defining characteristics of Treg cells is a high density of IL-2Rα chain, and in addition, because IL-2 is required for Treg cell survival and their suppressor capacity, TCR-activated Treg cells express fully functional high affinity trimeric IL-2Rs, capable of signaling, as well as IL-2 internalization and metabolism. Direct measurement revealed that compared with T_{eff} cells, Treg cells bound and internalized significantly more IL-2-Fc fusion protein, a result predictable based upon the relative IL-2R densities of Treg cells versus T_{eff} cells. In addition, to further explore the concept that Treg cells can cause T_{eff} cell death by depleting IL-2, anti-IL-2 was added to T_{eff} cells stimulated with anti-CD3/28; blocking IL-2 leads to progressive cell death, so that by day three, 40% are apoptotic.

As a final test of cytokine deprivation apoptosis as an explanatory mechanism fundamental to Treg-mediated suppression of T_{eff} cells, Bim-deficient responder T cells were tested for their susceptibility to Treg-mediated apoptosis. As already noted, the pro-apoptotic molecule Bim becomes activated upon cytokine withdrawal, and is pathognomonic for cytokine withdrawal apoptosis, thereby serving to distinguish this cause of cell death as compared with receptor-mediated cell death, such as that due to Fas or TNFα stimulation. Thus, in co-culture with Treg cells, Bim-deficient T_{eff} cells are completely protected from apoptosis and proliferate just as well as T_{eff} cells cultured with control cells, without Tregs present. Therefore, Bim-deficient T_{eff} cells are unresponsive to Treg cell-mediated suppression because they are capable of resisting cytokine deprivation apoptosis.

Despite these very convincing data generated *in vitro*, before accepting cytokine deprivation apoptosis as a mechanism responsible for Treg-mediated suppression, *in vivo* validation is necessary. Thus, Pandiyan and Lenardo examined the model of the generation of inflammatory bowel disease (IBD) by T_{eff} cells when transferred to lymphopenic *scid* mice. By taking advantage of Thy-1.1 and Thy-1.2 congenic mice, it was possible to transfer 4×10^5 Thy-1.1$^+$ T_{eff} cells to *scid* recipients followed one week later by either PBS or 1×10^6 Thy-1.2$^+$ Treg cells. Without the transfer of TReg cells, the recipient mice suffer from extensive colitis, manifested by elongated crypts, massive cell infiltration, thickened walls, and overall colonic shortening, indicative of "autoimmune colitis." However, mice that also received 2.5-fold more Tregs versus T_{eff} cells had only modest thickening of colon walls, but otherwise had normal luminal architecture, normal colon length and little or no evidence of autoimmune disease. Examination of the Thy-1.1$^+$ T_{eff} cells in the mesenteric lymph nodes and colons one week after transfer of the Treg cells revealed the presence of many apoptotic cells as monitored by Annexin V and TUNEL staining, indicating that Treg cells suppress antigen-activated T_{eff} cells *in vivo* by inducing their apoptosis. In support of this interpretation, there was no evidence for Treg suppression of early T_{eff} cell proliferation, monitored only two days after transfer. Moreover, transfer of Bim-deficient T_{eff} cells failed to undergo apoptosis.

The cytokine concentrations available to the cells not only regulate survival and proliferation of T_{eff} cells, but even the Tregs themselves. In a follow-up report, Pandiyan and Lenardo examined the fate of Treg cells themselves in relationship to the IL-2Rγ_c, which is also utilized by IL-4, IL-7, IL-9, IL-15, and IL-21.[8] They found that the other γ_c-utilizing cytokines can substitute for IL-2 in maintaining Treg cell viability, but that only IL-2 promotes Treg cell proliferation. Thus, *in vitro*, Treg cells (i.e. cells sorted to be 90–95% CD4$^+$CD25$^+$ using magnetic beads) activated via soluble anti-CD3/28 and examined after 72–96 hours revealed a dramatic Treg cell death of 75–90% of the cells, as indicated by flow cytometry forward scatter and propidium iodide (PI) staining. Moreover, addition of IL-2, and other γ_c-utilizing cytokines (IL-4, IL-7, IL-15, IL-21) rescued their viability, while IL-23, a non-γ_c-utilizing cytokine did not. Given the fact that freshly isolated Treg cells express IL-2Rα, as well as IL-4Rα, IL-7Rα, and IL-15Rα, as also do T_{eff} cells, it is particularly noteworthy that only IL-2 is capable of inducing Treg cell proliferation, and the other γ_c-utilizing cytokines cannot. Thus, if cells cannot produce IL-2, such as Treg cells, then they must depend upon T_{eff} cells for their IL-2, as the other γ_cR-utilizing cytokines will function to keep them alive, but will not signal rapid proliferation in response to antigenic signaling. It follows that in the absence of IL-2, Tregs could respond to other γ_cR-utilizing cytokines, which could induce both survival and slow, homeostatic proliferation of Treg cells in the periphery. Moreover, as shown by Pandiyan and Leonard, the other γ_cR-utilizing cytokines can maintain both CD25 expression and FOXP3 expression by Treg cells. Moreover, without FOXP3 expression, such as occurs in the scurfy mice or IPEX patients, the lack of functional FOXP3 precludes the generation of cells that cannot make IL-2 but can respond to it. As stated by Pandiyan and Lenardo, "along with chronic TCR stimulation via self-peptides, the availability of γ_cR-utilizing cytokines probably functions to keep the Treg number in constant check, thus maintaining both the protective and regulatory arms of the immune system in balance."

References

1. Thornton, A.M., and Shevach, E.M. (1998) CD4+CD25+ immunoregulatory T cells suppress polyclonal T cell activation *in vitro* by inhibiting interleukin 2 production. *J. Exp. Med.* **188**:287–296.
2. Takahashi, T., Kuniyasu, Y., Toda, M., Sakaguchi, N., Itoh, M., Iwata, M., Shimizu, J., and Sakaguchi, S. (1998) Immunologic self-tolerance maintained by CD25+CD4+ naturally anergic and suppressive T cells: induction of autoimmune disease by breaking their anergic/suppressive state. *Int. Immunol.* **10**:1969–1980.
3. Thornton, A., Piccirillo, C., and Shevach, E. (2004) Activation requirements for the induction of CD4+CD25+ T cell suppressor function. *Eur. J. Immunol.* **34**:366–376.
4. Thornton, A.M., Donovan, E.E., Piccirillo, C.A., and Shevach, E.M. (2004) Cutting edge: IL-2 is critically required for the *in vitro* activation of CD4(+)CD25(+) T cell suppressor function. *J. Immunol.* **172**:6519–6523.
5. de la Rosa, M., Rutz, S., Dorninger, H., and Scheffold, A. (2004) Interleukin-2 is essential for CD4+CD25+ regulatory T cell function. *Eur. J. Immunol.* **34**:2480–2488.
6. Scheffold, A., Huhn, J., and Hofer, T. (2005) Regulation of CD4+CD25+ regulatory T cell activity: it takes (IL-)two to tango. *Eur. J. Immunol.* **35**:1336–1341.
7. Pandiyan, P., Zheng, L., Ishihara, S., Reed, S., and Lenardo, M. (2007) CD4+CD25+FOXP3+ regulatory T cells induce cytokine deprivation-mediated apoptosis of effector CD4+ T cells. *Nat. Immunol.* **8**:1353–1362.
8. Pandiyan, P., and Lenardo, M. (2008) The control of CD4+CD25+FOXP3+ regulatory T cell survival. *Biol. Direct* **3**:1745–1757.

Chapter 21

FOXP3 Restricts But Does Not Suppress IL-2 Production

Given that FOXP3 expression in the thymus is purportedly important in the generation of nTreg cells, and in view of the fact that FOXP3 expression is inducible *in vitro* in mature peripheral T cells (iTregs) via activation by TCR/CD3/CD28 stimulation, the molecular pathways that control FOXP3 gene expression become relevant. Because the kinetics of FOXP3 expression after engagement of the TCR resembled the delayed kinetics of CTLA-4 expression, we conjectured that although the TCR complex (signal #1) might initiate FOXP3 gene expression, there might also be contributions from the co-stimulatory pathway (signal #2), as well as via the IL-2/IL-2R JAK1/3 STAT5 pathway (signal #3). In this regard, Ritz and colleagues first showed that IL-2 regulates FOXP3 expression in human CD4$^+$CD25$^+$ T cells through a STAT5-dependent mechanism,[1] and Farrar's group showed that the IL-2Rβ chain-dependent activation of STAT5 is necessary for the expression of murine nTreg FOXP3, thereby suggesting that either IL-2 or IL-15 are involved.[2] However, the molecular mechanisms responsible for expression of FOXP3 by mature, peripheral TCR-activated T cells, and the consequences of this expression for T cell function had not been examined in detail.

First, in order to establish a baseline from which to interpret our experiments, we monitored FOXP3 expression by freshly isolated PBMCs.[3] In agreement with the report by Gavin and co-workers,[4] a survey of 19 normal volunteers revealed that only ~2–3% of T cells express detectable FOXP3, with most expressed by CD4$^+$ T cells (2.6 ± 0.3%) versus CD8$^+$ T cells (0.2 ± 0.05%) (mean ± SEM).

Furthermore, as also reported by Gavin and co-workers, we found that only CD25$^+$ T cells express FOXP3, and that only ~10% of both CD4$^+$CD25$^+$ and CD8$^+$CD25$^+$ cells were FOXP3$^+$. Moreover, the levels of both CD25 and FOXP3 expression are low as detected by the fluorescent intensities. Thus, in the human there were no cells with "high CD25" and FOXP3 expression that might be assumed to be *bona fide* nTregs, based on murine data.

To examine the effect of T cell activation on T cell FOXP3 expression, we purposely performed our *in vitro* stimulations solely with soluble anti-CD3, as compared with using solid-phase antibody, and also we used whole PBMC populations, instead of using purified T cells. Thus, we relied on monocytes for presentation of the MoAbs, and for co-stimulation via B7 molecules, hoping to mimic the physiological activation via the APC pMHC-B7 molecular complex. We reasoned that because total human PBMC populations contain ~70% T cells, 10% monocytes, 10% B cells, and 10% NK cells, our first experiments should yield the induction of FOXP3 expression regardless of the molecules involved or their cellular origin. Also, by using total PBMCs, we avoided the potential criticism that our cell separation procedures might be inefficient and leave behind a residual subset of cells responsible for FOXP3 expression. In addition, by relying on monocytes to furnish the accessory molecule ligands, B7.1/2, we could avoid potential artifacts that could be introduced by using anti-CD28 for co-stimulation, which would be required if purified T cells were used.

With this experimental system, we found that IL-2Rα chain expression increased earlier than FOXP3 expression, first detectable by six hours after the addition of anti-CD3, increasing to >80% of total T cells by 24 hours after activation, and ~90% of T cells by 48 hours. Thus, the culture system using total PBMCs and activation with soluble anti-CD3 supported productive activation of most T cells, both CD4$^+$ and CD8$^+$. By comparison, FOXP3 expression lagged behind, with an increase above baseline first detectable only after 12–18 hours, and maximal increases after 24 hours. In addition, only 14.6 ± 0.6% (mean ± SEM, n = 29) of CD4$^+$ T cells were FOXP3$^+$ after 24 hours, while 8.9% ± 1.0% (mean ± SEM, n = 27) of

CD8⁺ T cells became FOXP3⁺. Accordingly, the stimuli necessary to activate the IL-2Rα chain expression versus FOXP3 expression appeared to be different, at least quantitatively if not qualitatively. However, the fact that both mature peripheral CD4⁺ T cells and CD8⁺ T cells are capable of FOXP3 expression effectively precludes the use of FOXP3 as a marker solely for nTreg cells, which mature in the murine thymus, and are confined to CD4⁺CD25⁺ T cells. It is also obvious that expression of FOXP3 by activation of mature peripheral human T cells via CD3 stimulation precludes the concept that FOXP3⁺ cells represent a separate and distinct differentiated T cell lineage, at least in the human. In this regard, it is noteworthy that although 80–90% of T cells become activated as evidenced by IL-2Rα chain expression, but only 10–15% become FOXP3⁺.

The timing of the anti-CD3-induced FOXP3 expression, which followed the expression of IL-2Rα chains, suggested that IL-2 itself might be involved in the induction of FOXP3 gene expression. The exogenous supplementation of IL-2 in 100-fold excess (i.e. 10 nM) of the concentration required to saturate high affinity IL-2Rs did not accelerate the expression of FOXP3, as would be expected if its expression depended upon IL-2R-derived signals. Moreover, exogenous IL-2 supplementation did not increase the proportion of cells that expressed FOXP3. However, the supplementation with exogenous IL-2 did prolong maximal FOXP3 expression beyond the first 24 hours, which also suggested that IL-2 might be involved in the activation of FOXP3 expression.

To test directly the hypothesis that TCR/CD3-induced expression of FOXP3 is mediated by signals derived from the IL-2R, MoAbs reactive with either the IL-2Rα chain, or with IL-2 itself were added to the cultures. Because low nM concentrations of these MoAbs are effective, this approach is perhaps the least disruptive to the cell population. Both of these MoAbs effectively blocked the TCR/CD3-induced expression of FOXP3, and excess exogenous IL-2 effectively competed for the MoAb suppression, thereby confirming their specificity.

Thus, TCR/CD3-induced FOXP3 expression is *ultimately* signaled via the IL-2R pathways, and both TCR as well as IL-2-derived signals appear to be involved. Inasmuch as FOXP3 serves as a negative regulator of IL-2 gene expression, the next obvious question became the effect of FOXP3 expression on IL-2 gene expression, not only by cells wherein FOXP3 is expressed, but also by cells cultured with the cells expressing FOXP3, especially as CD4+C25+FOXP3+ cells were reported to actively suppress IL-2 gene expression by CD4+CD25- T$_{eff}$ cells, as discussed earlier. However, when examined directly in the same unseparated cell population, it was readily apparent that IL-2 gene expression and FOXP3 gene expression are mutually exclusive, i.e. the cells that become FOXP3+ do not express IL-2, while the cells that are FOXP3- do express IL-2. To test this observation further, PBMCs were activated via anti-CD3 for 24 hours to maximize FOXP3 expression, and then re-stimulated for just six hours with anti-CD3 + anti-CD28 in the presence of Brefeldin-A to block IL-2 secretion. In this experimental design, the cells are present in a physiological setting, at relative concentrations that should allow detection of "active" suppression of IL-2 expression by the FOXP3+ cells, should such a phenomenon actually occur. As shown by a

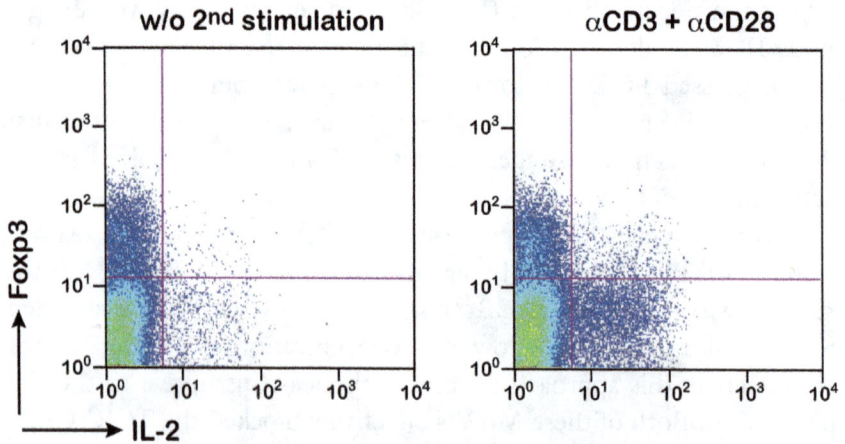

Figure 21.1:   **FOXP3 restricts but does not suppress IL-2 expression.** Human PBMCs were activated with anti-CD3 (10 mg/mL) + IL-2 (1 nM) for 24 hours to induce FOXP3 expression, then harvested, washed and re-stimulated with anti-CD3 + anti-CD28 for six hours in the presence of Brefeldin-A. Plots show CD4+ T cells reacted with FOXP3 versus IL-2 MoAbs. (From Popmihajlov, Z. and Smith, K.A. 2008. *PLoS ONE* **3**:e1581.)

representative experiment in Fig. 21.1, very few FOXP3$^+$ cells expressed IL-2, whereas almost all of the IL-2 expressed came from FOXP3$^-$ cells. Thus, IL-2 expression is restricted in FOXP3$^+$ cells but these cells do not suppress IL-2 expression by the FOXP3$^-$ cells in the culture. Moreover, compilation of data from five separate experiments revealed that after anti-CD3 restimulation, the mean IL-2 expression by FOXP3$^-$ cells increased 10-fold. Accordingly, in this physiological experimental setting, with a ratio of FOXP3 $^+/^-$ cells of 1:5, it is obvious that FOXP3 expression restricts the expression of IL-2, but FOXP3$^+$ cells do not actively suppress the expression of IL-2 by FOXP3$^-$ cells.

References

1. Zorn, E., Nelson, E., Mohseni, M., Porcheray, F., Kim, H., Litsa, D., Belluci, R., Raderschall, E., Canning, C., Soiffer, R.J., *et al.* (2006) IL-2 regulates FOXP3 expression in human CD4$^+$CD25$^+$ regulatory T cells through a STAT-dependent mechanism and induces the expansion of these cells *in vivo*. *Blood* **108**:1571–1579.
2. Burchill, M., Yang, J., Vogtenhuber, C., Blazar, B., and Farrar, M. (2007) IL-2 receptor beta-dependent STAT5 activation is required for the development of FOXP3$^+$ regulatory T cells. *J. Immunol.* **178**:280–290.
3. Popmihajlov, Z., and Smith, K. (2008) Negative feedback regulation of T cells via interleukin-2 and FOXP3 reciprocity. *PLoS ONE* **3**:e1581.
4. Gavin, M., Torgerson, T., Houston, E., DeRoos, P., Ho, H., Stray-Pedersen, A., Ocheltree, E., Greenberg, P., Ochs, H., and Rudensky, A. (2006) Single-cell analysis of normal and FOXP3-mutant human T cells: FOXP3 expression without regulatory T cell development. *PNAS* **103**:6659–6664.

Chapter 22

Both the TCR and IL-2 Regulate FOXP3 Expression

The experiments detailed above are important, in that they show that within a mixed cell population, if one blocks the IL-2/IL-2R interaction, TCR triggering together with any other cytokines already present *in situ* or produced by any of the cells are insufficient to induce FOXP3 expression. However, these data do not adequately delineate the signaling pathways or their roles in actually promoting FOXP3 expression. Utilizing either human or murine T cells, it has been shown that purified mature peripheral CD4$^+$CD25$^-$FOXP3$^-$ T cells can be made to yield 100% FOXP3$^+$ cells provided they are stimulated by solid-phase anti-CD3/28 and cultured for several days in IL-2 + TGFβ.[1–3] Also, in these experiments, it is critical to use solid-phase anti-CD3, as soluble MoAb fails to activate FOXP3 expression in murine cells. This is particularly noteworthy, in that solid-phase anti-CD3 has always been considered to produce a very "strong" signal. Based on the hypothesis that the TCR is a mechanoreceptor and is triggered by torque forces, the use of solid phase anti-CD3 may well provide for maximum torque. In addition, the exogenous addition of TGFβ is just as critical for FOXP3 expression by murine T cells.[4] However, if one uses T cells from IL-2 (-/-) mice they do not become FOXP3$^+$ unless IL-2 is supplied exogenously, clearly emphasizing a critical role played by IL-2.[3]

Since the expression of IL-2 and IL-2Rα chains is TCR/CD3$^-$ activation dependent, it is difficult to dissect the roles played by the TCR versus the IL-2R in triggering the expression of FOXP3. However, there have been several recent reports that focus on the mechanism(s) regulating FOXP3 gene expression. As already detailed, the TCR/CD3/CD28 signaling complex activates three families of transcription factors, Rel, AP-1, and NFAT, which then cooperate to activate IL-2 gene transcription.[5,6] Mantel and co-workers explored the structure and function of the human FOXP3 promoter and provided evidence that there is a proximal promoter localized between −511/+176 bp, and that this element has at least three NFAT REs and three AP-1 REs, thereby implicating TCR/CD3 regulation of FOXP3 expression. In addition, others have identified a TCR-responsive RE in the first intron of the FOXP3 gene that is dependent on a cAMP RE-binding protein/activating transcription factor site.[7]

Whatever the role of TCR/CD3/CD28-triggered Rel, AP-1 and NFAT or CREB in regulating the expression of FOXP3, there have now been several reports that have found STAT5 to be a key TF that regulates the FOXP3 gene. Thus, Michael Farrar's group showed convincingly that STAT5 is both necessary and sufficient for FOXP3 expression, and that there are six potential STAT5 REs in the gene, three in the promoter region, and three in the first intron.[8] Moreover, using chromatin immunoprecipitation (ChIP) assays, evidence was presented that pSTAT5 binds to the FOXP3 gene promoter *in vivo*. In these experiments, attention focused on the generation of nTregs in the thymus, and it was concluded that the IL-2Rβ chain is key, and that either IL-2 or IL-15 can trigger the expression of FOXP3 in the thymus, but that IL-2 is critical in the periphery to maintain FOXP3 expression after the cells have exited the thymus. Thus, IL-2 (-/-) T cells exit the thymus with near normal numbers of FOXP3+ cells, but then over time FOXP3 expression is gradually lost, eventually resulting in multi-organ polyclonal T cell-mediated autoimmunity.

In this regard, the present immunological dogma is that nTregs are generated in the thymus in response to TCR/CD3 signaling by autologous pMHC complexes, which are thought to trigger intermediate affinity TCRs, i.e. stronger than those necessary for positive selection, but weaker than those causing negative selection. If true, then these same autologous pMHC complexes must be present in the periphery, serving to chronically and tonically trigger self-reactive TCRs, which promote FOXP3 expression together with pSTAT5 signals derived from IL-15 and/or IL-2. That STAT5 activation is both necessary and sufficient for FOXP3 expression was shown convincingly by John O'Shea's team, which reported that STAT3 and STAT5 have opposing roles in the regulation of FOXP3.[9] To determine the role of pSTAT5 in lymphopoiesis, STAT5a/b doubly deleted mice were generated. However, because only the first exons of these genes were deleted, and the rest of the coding region was in-frame, these mice still expressed residual STAT5 protein. Consequently, there was reduced but not absent thymic and peripheral FOXP3 expression.

Truly STAT5 (-/-) mice, in which the entire STAT5a/b loci are deleted, die perinatally due to anemia, but a ~2% of newborns survive for six to eight weeks. Examination of these mice revealed marked reductions in both thymic and peripheral IL-2Rα chain[+] and FOXP3[+] T cells, comparable to the reductions observed in JAK3 (-/-) and IL-2Rγ_c (-/-)[Y] mice. Thus, signaling via the IL-2Rβ/γ chains, activating JAK1/3 and pSTAT5 is obligatory for the expression of FOXP3. In this regard, it is noteworthy that IL-4, which activates STAT6, cannot substitute for the IL-2Rβ/γ chain activation of STAT5, while IL-6, which activates STAT3, actually suppresses FOXP3 expression. The reciprocal regulation of FOXP3 by STAT5 versus STAT3 is of interest with regard to the discrimination of self versus non-self pMHC complexes, in that IL-6, a cardinal pro-inflammatory cytokine and a prime product of macrophages, may well function as a molecular switch to alert the immune system to foreign antigens, thereby promoting activation and suppressing suppression.

If there is a deficiency of IL-2/pSTAT5/FOXP3 activation as in the IL-2 (-/-) or IL-2Rα (-/-) or IL-2Rβ (-/-), it is likely that IL-6 and IL-21, both of which activate STAT3, would predominate. Also, as STAT3 has been implicated in the activation of IL-17 expression, which activates APCs to produce even greater amounts of pro-inflammatory cytokines like IL-6 and TNFα, a relative deficiency of IL-2 could very well account for the accumulation of myeloid cells and polymorphonuclear leukocytes that is typically seen in the IL-2 and the IL-2R (-/-) mice.

The most important conclusion, given all of these data, is that FOXP3 expression is both TCR and IL-2-dependent, especially when expressed in the periphery by mature CD4$^+$ T cells. In this regard, it is noteworthy that not all human T cells express FOXP3 when activated polyclonally via anti-CD3, even when excess IL-2 is supplied exogenously. Thus, perhaps only selected cells can differentiate to FOXP3$^+$ cells, or we need to further define the parameters that lead to FOXP3 expression.

References

1. Allan, S., Crome, S., Crellin, N., Passerini, L., Steiner, T., Bacchetta, R., Roncarolo, M., and Levings, M. (2007) Activation-induced FOXP3 in human T effector cells does not suppress proliferation or cytokine production. *Int. Immunol.* **19**:345–354.
2. Tran, D.Q., Ramsey, H., and Shevach, E.M. (2007) Induction of FOXP3 expression in naive human CD4$^+$FOXP3 T cells by T-cell receptor stimulation is transforming growth factor-{beta} dependent but does not confer a regulatory phenotype. *Blood* **110**:2983–2990.
3. Davidson, T., Dipaolo, R., Anderson, J., and Shevach, E. (2007) IL-2 is essential for TGF-beta-mediated induction of FOXP3$^+$ T regulatory cells. *J. Immunol.* **178**:4022–4026.
4. Chen, W., Jin, W., Hardegen, N., Li, K., Li, L., Marinos, N., McGrady, G., and Wahl, S. (2003) Conversion of peripheral CD4$^+$CD25$^-$ naive T cells to CD4$^+$CD25$^+$ regulatory T cells by TGF-beta induction of transcription factor FOXP3. *J. Exp. Med.* **198**:1875–1886.
5. Garrity, P.A., Chen, D., Rothenberg, E.V., and Wold, B.J. (1994) Interleukin-2 transcription is regulated *in vivo* at the level of coordinated binding of both constitutive and regulated factors. *Mol. Cell. Bio.* **14**:2159–2169.
6. Rothenberg, E.V., and Ward, S.B. (1996) A dynamic assembly of diverse transcription factors integrates activation and cell-type information for interleukin-2 gene regulation. *Proc. Natl. Acad. Sci. USA* **93**:9358–9365.
7. Kim, H.-P., and Leonard, W. (2007) CREB/ATF-dependent T cell receptor-induced *FOXP3* gene expression: a role for DNA methylation. *J. Exp. Med.* **204**:1543–1551.
8. Burchill, M., Yang, J., Vogtenhuber, C., Blazar, B., and Farrar, M. (2007) IL-2 receptor beta-dependent STAT5 activation is required for the development of FOXP3$^+$ regulatory T cells. *J. Immunol.* **178**:280–290.
9. Yao, Z., Kanno, Y., Kerenyi, M., Stephens, G., Durant, L., Watford, W., Laurence, A., Robinson, G., Shevach, E.M., Morrigl, R., *et al.* (2007) Nonredundant roles for STAT5a/b in directly regulating FOXP3. *Blood* **109**:4368–4375.

Chapter 23

The Effects of FOXP3 Expression

As FOXP3 expression is controlled by the TCR and IL-2, and therefore restricted to T cells, the effects of FOXP3 expression on T cells is the next obvious question. Two functions have been attributed to FOXP3:

1) the restriction of cytokine gene expression by FOXP3$^+$ cells, thereby resulting in energy, and
2) the FOXP3$^+$ T cell "active suppression" of FOXP$^-$ cell cytokine production, especially IL-2, resulting in their diminished proliferation.

As demonstrated, the expression of FOXP3 and IL-2 is mutually exclusive, so that it appears that IL-2 induces its own inhibitor, a prerequisite for a classic internal hormonal control system.

Soon after the FOXP3 gene was identified, transgenic mice were constructed using the entire coding region of the gene along with its regulatory elements.[1] Thus, over-expression of FOXP3 due to gene amplification was achieved. The results were very informative, in that the only abnormalities found were within peripheral T cells. Thymocyte maturation, as well as positive and negative selection of T cells within the thymus appeared normal. However, there was a marked diminution of peripheral mature T cells, both CD4$^+$ and CD8$^+$ T cells. In addition, there were marked functional deficiencies, manifest especially by markedly depressed production of IL-2, and consequently of TCR/CD3-induced proliferation, as well as the generation of cytolytic T cells. Exogenous IL-2 supplementation circumvented most of these defects, thereby attesting to the primary role of IL-2 gene expression and function in the generation of T cell immune responsiveness, as well as the function of FOXP3 to restrict IL-2 gene expression.

As already detailed, the loss of FOXP3 gene expression, as occurs in the scurfy mouse and the IPEX individuals, results in a marked over-expression of IL-2 and other cytokine genes by T cells, while the remainder of somatic cells appears normal. Moreover, triggering via the TCR is enhanced, so that co-stimulation via CD28 becomes unnecessary. Thus, these data are consistent with a lowered activation threshold in triggering of cytokine gene expression, so that one would speculate, based upon the Quantal Theory, that a lower number of triggered TCRs is sufficient to activate the expression of IL-2 and other cytokine genes when FOXP3 is missing. Accordingly, these data indicate how signals emanating at the membrane can be attenuated at the level of promoters.

Regarding the molecular mechanism(s) of FOXP3 action, initial data indicated that there are several forkhead-specific REs adjacent to critical NF-AT REs in the promoters of several cytokine genes.[2] Moreover, the effect of FOXP3 expression versus FOXP1 and FOXP2 was examined, and only FOXP3 specifically inhibited cytokine gene expression (i.e. IL-2, IL-4, and IFNγ) by primary murine CD4[+] T cells. In subsequent experiments, FOXP3 was found to inhibit Rel family members (Rel-A and NF-AT) from the transcriptional activation of cytokine genes without inhibiting their DNA binding capacities.[3] Accordingly, these data implied that FOXP3 functions by interacting directly with and inhibiting key TFs from activating the transcription of cytokine genes. This notion was supported further by the crystal structure of FOXP2 complexed with NF-AT; there is a large protein-protein interaction interface between these two proteins.[4] In addition, point mutations of FOXP3 predicted from this crystal structure to disrupt the FOXP2/NF-AT complex correlated with the loss of FOXP3 inhibition of IL-2 gene expression.

With regard to the effects of the FOXP3-inhibition of cytokine gene expression, one must be cognizant of both primary as well as secondary effects, since cytokines have their own immediate/early (i.e. direct) and downstream genes that they regulate.[5] With regard to FOXP3 target genes, the combination of ChIP analysis with DNA arrays has indicated that FOXP3 plays a dual role as both a transcriptional activator, as well as a transcriptional repressor.[6,7] These approaches have shown that many FOXP3-regulated genes encode proteins associated with the TCR signaling pathways, and most have shown suppressed activation when the cells were TCR stimulated.

In this regard, recent data using reporter cells that express either WT FOXP3 or mutant inactive FOXP3 have shown that one of the important genes involved in TCR-activated IL-2 gene expression is NF-AT2.[8] NF-AT2 is typically expressed at low levels in resting T cells, but is transcriptionally activated by TCR-signaled activation of NF-AT1. FOXP3 competes with NF-AT1 for binding to the NF-AT2 promoter, which is important for the sustained anergic phenotype of FOXP3$^+$ cells. The data suggest that FOXP3 functions not only to suppress the first wave of NF-AT-mediated TCR-activated cytokines, but also to guarantee cytokine gene repression by influencing secondary genes as well.

By comparison to these mounting data supporting a role for FOXP3 in T cells that express this novel transcriptional regulator, data suggesting that FOXP3$^+$ cells can actually "actively" suppress T_{eff} cells have not been as forthcoming. One intriguing report from Silvia Deaglio and co-workers has provided a plausible mechanism whereby FOXP3$^+$ cells could "actively" suppress FOXP3$^-$ T_{eff} cells polyclonally.[9] According to their report, T-Reg cells express two ectoenzymes on their cell surface, CD39 and CD73, which together generate pericellular adenosine from extracellular nucleotides. Moreover, T_{eff} cells express 7-transmembrane cell surface adenosine receptors, which when triggered by adenosine fabricated by T-Reg cells, generate intracellular cAMP. Deaglio's report did not offer a molecular mechanism of suppression, but elevated intracellular cAMP is well known to inhibit both IL-2 production and IL-2-signaled T cell proliferation. This mechanism also satisfies the requirement that T-Reg cells are capable of suppressing T_{eff} cells in an antigen nonspecific manner.[10] Another recent report invokes the transference of cAMP from T-Regs to T_{eff} cells via gap junctions.[11]

Another recent report from Shevach's group has suggested that T-Regs do not act directly on T_{eff} cells, but rather function to prevent up-regulation of the co-stimulatory molecules CD80/86 (B7s) on DCs.[12] Using human T-Reg cells with mouse DCs and mouse T_{eff} cells, they found that LFA-1 on the T-Reg cells interacts with ICAM-1 on DCs, and that blocking this interaction circumvents the suppression exerted by T-Regs. Although they did not show it directly, these investigators implied that the interaction between the T-Regs and the DCs suppresses T_{eff} cell IL-2 production by damp-ening their activation via the APCs, thereby inhibiting the T cell proliferation. Moreover, they purport that IL-2 consumption plays no role in T-Reg-mediated suppression, because mouse T-Regs did not appreciably suppress human T_{eff} cells. Since mouse IL-2Rs can bind, internalize and degrade human IL-2, these investigators inter-preted the lack of suppressive function of mT-Regs on hT_{eff} cell proliferation as evidence that the mT-Regs could not possibly be consuming hIL-2. They also showed that a MoAb reactive with the human IL-2Rα chain expressed by human T-Regs failed to block their suppressive effects on murine T_{eff} cell proliferation, thus furthering their case that IL-2 consumption is not involved in T-Reg-mediated suppression.

References

1. Khattri, R., Kasprowicz, D., Cox, T., Mortrud, M., Appleby, M., Brunkow, M., Ziegler, S., and Ramsdell, F. (2001) The amount of scurfin protein determines peripheral T cell number and responsiveness. *J. Immunol.* **167**:6312–6320.
2. Schubert, L., Jeffrey, E., Zhang, Y., Ranmsdell, F., and Ziegler, S. (2001) Scurfin (FOXP3) acts as a repressor of transcription and regulates T cell activa-tion. *J. Biol. Chem.* **276**:37672–37679.
3. Bettelli, E., Dastrange, M., and Oukka, M. (2005) FOXP3 interacts with nuclear factor of activated T cells and NF-kB to repress cytokine gene expression and effector functions of T cells. *PNAS* **102**:5138–5143.
4. Wu, Y., Borde, M., Heissmeyer, V., Feuerer, M., Lapan, A., Stroud, J., Bates, D., Guo, L., Han, A., Ziegler, S., *et al.* (2006) FOXP3 controls regulatory T cell function through cooperation with NFAT. *Cell* **126**:375–387.
5. Beadling, C., and Smith, K. (2002) DNA array analysis of interleukin-2-regulated immediate/early genes. *Med. Immunol.* **1**:2.

6. Marson, A., Kretschmer, K., Frampton, G., Jacobson, E., Polansky, J., MacIsaac, K., Levine, S., Fraenkel, E., von Boehmer, H., and Young, R. (2007) FOXP3 occupancy and regulation of key target genes during T cell stimulation. *Nature* **455**:931–935.

7. Zheng, Y., Josefowicz, S., Kas, A., Chu, T.-T., Gavin, M., and Rudensky, A. (2007) Genome-wide analysis of FOXP3 target genes in developing and mature regulatory T cells. *Nature* **455**:936–940.

8. Torgerson, T., Genin, A., Chen, C., Zhang, M., Zhou, B., Anover-Sombke, S., Frank, M., Dozmorov, I., Ocheltree, E., Kulmala, P., *et al.* (2009) FOXP3 inhibits activation-induced NFAT2 expression in T cells thereby limiting effector cytokine expression. *J. Immunol.* **183**:907–915.

9. Deaglio, S., Dwyer, K.M., Gao, W., Friedman, D., Usheva, A., Erat, A., Chen, J.-F., Enjyoji, K., Linden, J., Oukka, M., *et al.* (2007) Adenosine generation catalyzed by CD39 and CD73 expressed on regulatory T cells mediates immune suppression. *J. Exp. Med.* **204**:1257–1265.

10. Thornton, A.M., and Shevach, E.M. (1998) CD4⁺CD25⁺ immunoregulatory T cells suppress polyclonal T cell activation *in vitro* by inhibiting interleukin-2 production. *J. Exp. Med.* **188**:287–296.

11. Bopp, T., Becker, C., Klein, M., Klein-Hessling, S., Palmetshofer, A., Serfling, E., Heib, V., Becker, M., Kubach, J., Schmitt, S., *et al.* (2007) Cyclic adenosine monophosphate is a key component of regulatory T cell mediated suppression. *J. Exp. Med.* **204**:1303–1310.

12. Tran, Q., Glass, D., Uzel, G., Darnell, D., Spalding, C., Holland, S., and Shevach, E. (2009) Analysis of adhesion molecules, target cells, and the role of IL-2 in human FOXP3⁺ regulatory cell suppressor function. *J. Immunol.* **182**:2929–2938.

Chapter 24

The Role of IL-2 in the Generation of Immune Responses *In Vivo*

At this time, it has become dogma that Treg cells "actively suppress" the immune response, although the mechanism(s) whereby these cells effect their suppression remain elusive. As already detailed, some investigators maintain that Tregs suppress the expression of the B7 co-stimulatory molecules on APCs by a cell contact-dependent mechanism that remains to be elucidated at the molecular level. However, this lack of effective APC co-stimulation of antigen-activated T_{eff} cells retards their production of IL-2, thereby attenuating their clonal proliferative expansion, the hallmark of a systemic immune response, as well as the hallmark of Treg suppression of T_{eff} cells.[1] Alternatively, the Treg "passive" binding, internalization and degradation of IL-2 produced by T_{eff} cells may well account for much, if not all, of the suppression of T_{eff} cell proliferation.[2,3] Of course, these two proposed mechanisms are not necessarily mutually exclusive.

Soon after the IL-2 (-/-) mouse was reported to be capable of a "normal" immune response to viral pathogens,[4] a new technique to monitor cell division both *in vitro* and *in vivo* was reported.[5] By using the dye 5-(and 6)-carboxyfluorescein diacetate succinimidyl ester (CFSE), it was shown for the first time in 1994 by Lyons and Parish that it is possible to monitor successive cell divisions. This technique was subsequently popularized by Barbara Fazekas de St. Groth and her colleagues, in that this new method when combined with adoptive transfer of lymphocytes, allowed one to study immune responses *in vivo*.[6,7]

185

Using this new technique, and following up on the observations by Horak and collaborators that T cell proliferative responses of cells from IL-2 (-/-) mice were only reduced by ~2/3 both *in vitro* and *in vivo* compared with WT mice,[4,8] D'Souza and Lefrancois reported a decade later that:

> "our results *conclusively demonstrated* that initial division of Ag-specific CD8+ T cells following priming was *IL-2 independent*. In contrast, the latter stage of the proliferative phase was IL-2-dependent."[9]

However, a close examination of the data underlying these claims suffers from assumptions made by these investigators. A mixture of equal numbers of CFSE-labeled ovalbumin (OVA)-reactive TCR-Tg (OT-1-RAG(-/-)) and OT-1-CD25(-/-)RAG(-/-) were transferred into naive recipients that were immunized a day later with vesicular stomatitis virus (VSV) containing *OVA*. They then monitored T cells at an "early time point" post-infection (57 hrs = 2.4 days), and:

> "found no obvious difference in the CFSE profile between the WT TCR-Tg and the CD25 (-/-) TCR-Tg within the LN, spleen and lung, thus revealing that *the initiation of cell division proceeded normally in the absence of IL-2-mediated signaling.*"

These investigators stated that IL-2Rα chain (-/-) cells are incapable of responding to IL-2, having made the assumption that lack of expression of the IL-2Rα chain precludes signaling via the IL-2Rβ and IL-2Rγ_c chains. As already detailed, cells that express only the IL-2Rβ and IL-2Rγ_c chains, can still bind IL-2, albeit with a 100-fold lower affinity (K_D = 1 nM), and because the IL-2Rβ and IL-2Rγ_c chains are responsible for signaling, one cannot assume that they cannot signal, particularly without showing any data to support such a claim. Moreover, they did not measure the IL-2 concentrations *in vivo*. Therefore, they cannot exclude that IL-2 concentrations high enough to bind to and trigger the intermediate affinity IL-2Rβ, γ_c occurred. Inasmuch as Jon Sprent had demonstrated so distinctly five years earlier[10] that both naive and memory phenotype CD8+ T cells express both IL-2Rβ and IL-2Rγ_c chains as parts of the IL-15R complex, such an assumption is clearly specious. Moreover, Tsunobuchi and co-workers reported that:

"serum IL-2 but not IL-15, was spontaneously detected in naive IL-2Rα (-/-) mice ... suggesting that spontaneous IL-2 production (or the lack of efficient consumption) may be responsible for expansion of the memory phenotype CD8⁺ T cells in naive IL-2Rα (-/-) mice."[11]

In other experiments by D'Souza and Lefrancois, OVA-TCR-Tg T cells on an IL-2 (-/-) background or an IL-2Rα (-/-) background were transferred into Tg mice that express OVA specifically within intestinal epithelial cells. By day six post-transfer of WT-OVA-Tg T cells, an impressive accumulation of the WT-OVA-Tg T cells into the epithelium occurred. By comparison, the transfer of the OVA-Tg T cells on an IL-2 (-/-) background resulted in a four- to five-fold decreased (i.e. 75%–80% decrease) accumulation. Also, the same OVA-Tg T cells on an IL-2Rα chain (-/-) background was even lower (six- to eight-fold or 85–88% decrease). These investigators did find a necessity for IL-2 for later stages of antigen-activated T cell proliferation, so that they concluded that "activated T cells initially undergo IL-2 independent proliferation, but reach a critical juncture where the requirement for IL-2 as a growth factor gains prominence." The authors speculated that either the early phases of CD8⁺ T cell proliferation might occur in the absence of any cytokines, or alternatively might be driven by other cytokines that utilize the IL-2Rγ chain, or perhaps even unknown cytokines might be responsible. This report was influential in furthering the conjecture that IL-2 is unnecessary for the initial proliferative response to an antigen, but is important for the subsequent expansion, especially in non-lymphoid tissues, where these other putative cytokines might be lacking.

This topic was addressed by Michael Bevan's group in a report that appeared in 2006. Hoping to circumvent the problems inherent in the IL-2 (-/-) and IL-2R (-/-) mice, which slowly accumulate activated T cells and suffer from autoimmune disorders, these investigators generated WT/IL-2Rα (-/-) mixed bone marrow chimeras. Thus, they transferred a 1:1 mixture of WT B6 and IL-2Rα

(-/-)-deficient bone marrow to WT B6 mice that had received 1000 rads of irradiation the day before. Since the irradiation rendered the recipient mice lymphopenic, the transferred cells then responded to the homeostatic proliferative stimuli from IL-7 and IL-15, and eight weeks later, the reconstituted mice had a 2:1 ratio of WT to IL-2Rα (-/-) CD4$^+$ T cells. The WT CD4$^+$ T cells were 17% + for IL-2Rα chain expression, and therefore had a "normal" complement of nTregs. These mice were then infected with LCMV and monitored for their CD8$^+$ T cell immune responses to the immunodominant epitope from the LCMV glycoprotein (GP$_{33-41}$). Consistent with the results of Lefrancois, this approach was purported to show little difference between the WT and the IL-2Rα (-/-) CD8$^+$ peptide-specific T cells, monitored by short-term peptide stimulated IFNγ production, when assayed at the height of the primary immune response on day 8 (15% decrease), or after the contraction phase (day 40; 30% decrease). However, by day 150 the IL-2Rα (-/-) CD8$^+$ T cells were only ~50% compared with WT. Thus, the (-/-) CD8$^+$ T cells were not maintained to the same level as were the WT cells. More striking was the secondary response. When rechallenged with the GP$_{33-41}$ epitope, WT CD8$^+$ T cells expanded more than 40-fold, while the IL-2Rα (-/-) CD8$^+$ T cells only expanded four-fold. This discrepancy was interpreted as not due to a lack of sufficient numbers of IL-2Rα (-/-) memory T cells prior to challenge, but instead due to a deficiency of an IL-2-programmed differentiation process necessary during the primary response that "renders the formation of fully responsive CD8$^+$ memory T cells capable of generating robust recall responses." Exactly what underlies this IL-2-driven differentiation process was left to future experiments.

The Bevan group interpreted their results with the assumption, like the Lefrancois team, that the IL-2Rα (-/-) T cells were "lacking the high affinity IL-2 receptor." Of course this is entirely correct. These cells cannot express high affinity trimeric IL-2Rs, because they cannot express the IL-2Rα chain. However, they can express intermediate affinity dimeric IL-2Rβ, γ_c chains, which are capable of signaling. Consequently, the data in this report are uninterpretable from the standpoint of IL-2 signaling. Another measure of IL-2 signaling, other than the numbers of T cells, would be necessary to exclude the possibility that the IL-2Rα (-/-) T cells did not receive IL-2 signals.

In 2002, Omar Perez and Garry Nolan first reported techniques that enabled the analysis of intracellular signaling molecules at the single cell level using the flow cytometer.[12] Because this technique allows one to supercede experiments at the population level and actually quantify intracellular signaling molecules and to trace the dynamics of signaling in individual cells, it promises to revolutionize our understanding of how signaling occurs.[13]

Long and Adler used this new technique to explore the immune reactivity of CD4[+] T cells in response to a virus infection.[14] To focus on antigen-specific T cells, TCR-Tg T cells that recognize an influenza virus hemagglutinin (HA) epitope were adoptively transferred into non-transgenic recipients that had been immunized with a recombinant vaccinia virus expressing HA. At four-hour intervals splenocytes were removed, fixed and immediately examined for expression of IL-2, IL-2Rα chain (CD25), and pSTAT5. As shown in Fig. 24.1, IL-2 expression was detectable after four hours and peaked at 8–12 hours, and then subsided and became barely detectable by 20–4 hours. Of interest, none of the cells that became IL-2[+] expressed pSTAT5, but only the IL-2[−] cells became pSTAT5[+], lagging behind IL-2 expression, peaking at 16 hours, then slowly subsiding. By comparison, the expression of the IL-2Rα chain (CD25) by the HA-specific T cells peaked a little earlier than pSTAT5, at 12–16 hours, as would be expected.

Also, of special interest, the endogenous normal CD4[+] T cells, of which 13% expressed IL-2Rα chain (CD25), at t = 0, ~15% of these cells already expressed pSTAT5, presumably because they were activated by self-peptide MHC interaction and also by IL-2. By four hours, when the HA-specific cells first became detectable for IL-2, the endogenous IL-2Rα chain (CD25)[+] cells all became pSTAT5[+], and remained pSTAT5[+] until after 20 hours. In other experiments, it was clearly shown that IL-2 expression and pSTAT5 expression were mutually exclusive, thereby indicating that IL-2[+] cells do not respond to the IL-2 that they themselves secrete. Instead, all

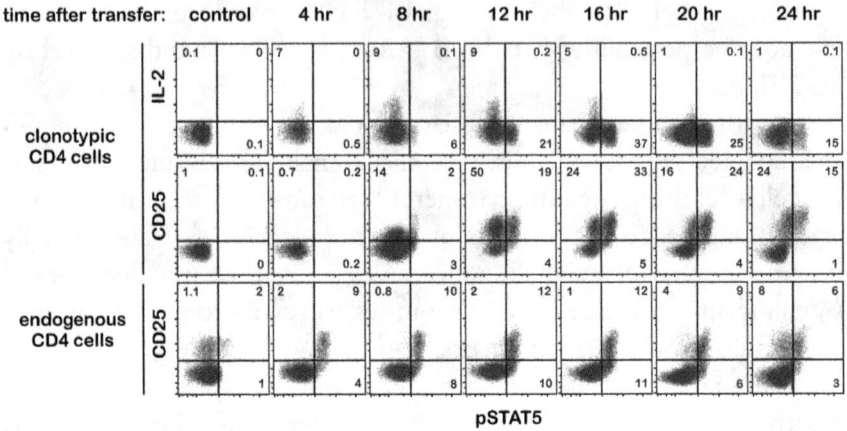

Figure 24.1:   **Time course of early IL-2 expression and signaling in antiviral CD4+ T cells.** Naïve Thy1.1+HA-specific TCR-Tg CD4+ T cells were adoptively transferred to vaccinia-HA-infected Thy1.2+ non-transgenic recipients, recovered from spleens at the indicated times and directly stained for IL-2, IL-2Rα chain (CD25), and pSTAT5. The % of positively expressing TCR-Tg cells (CD4+Thy1.1+) are indicated. IL-2Rα chain (CD25) versus pSTAT5 staining on endogenous (Thy1.1−) CD4+ T cells is also shown. Non-infected adoptive transfer recipients served as controls. (From: Long, M. and Adler, A.J. 2006. *J. Immunol.* **177**:4257–4261.)

of the IL-2-responsiveness, as determined by the phosphorylation of STAT5 occurs in a paracrine mode.

This report was recently followed up by a more extensive analysis of the phenomena by Abul Abbas' group.[15] Also using a TCR Tg model system (DO11.10) of CD4+ T cells, but specific for a chicken ovalbumin (OVA) peptide, these investigators sought to define the cells that respond to IL-2 as well as the time-course of the response *in vivo*. The DO11 T cells were adoptively transferred to normal syngeneic BALB/c recipients and then immunized with OVA or injected with PBS as a control. Monitoring pSTAT5 expression by flow cytometry at early time points after the injections, it was readily apparent that the DO11 TCR-Tg T cells begin producing IL-2 as early as two hours after immunization and stopped producing it by 24 hours, thus demonstrating that the transient expression of IL-2 first noted *in vitro*,[16] also occurs *in vivo*, and is consistent with earlier data from Singh and co-workers.[17]

As for the IL-2 responders, as early as six hours after the immunization, there was a marked increase in pSTAT5 in all endogenous phenotypic Treg cells (i.e. CD4+CD25+FOXP3+), but not in phenotypic endogenous naive CD4+ or CD8+ T cells, which do express detectable IL-2Rα chains.[10] Furthermore, proof that IL-2 produced by the antigen-stimulated TCR-Tg cells was responsible for the induction of pSTAT5 by the Tregs, was obtained by the use of TCR-Tg cells with their IL-2 genes deleted; these cells did not induce Treg pSTAT5. This is an extremely important finding, because it indicates that IL-2 is the sole cytokine involved in the immunological activation of Treg cells *in vivo*. Given these observations it is also noteworthy that ~15% of normal Tregs are positive for pSTAT5 in the absence of immunization, a finding that is also consistent with the data of Long and Adler, and also from previous reports,[18,19] which showed that the maintenance of Treg cells depends upon low levels of IL-2 present in the environment.

Also of considerable interest, examination of the transferred antigen-stimulated TCR-Tg cells at the same time intervals when the endogenous Treg cells were all pSTAT5+ (i.e. 6, 12, and 24 hours), none of the TCR-Tg cells were pSTAT5+. This lack of reactivity could not be attributed to the lack of expression of IL-2Rs, in that all three chains could be detected. However, the IL-2Rα chain density was not as great as it was on endogenous Treg cells, which ranged a half log greater density at the same time interval (12 hours). Thus, as discussed in Chap. 9, as the IL-2Rα chain density increases, the affinity of IL-2 binding increases, thereby increasing the efficiency of IL-2 binding, internalization and degradation. Also, the naive TCR transgenic T cells may not be fully competent to receive IL-2 signals, because JAK3 must be expressed in response to TCR-generated signals.[20] Consistent with this interpretation, if a secondary antigenic stimulus was administered 2.5 days after antigen priming, then the TCR-TG cells expressed pSTAT5 within eight hours. Finally, very similar data were obtained when the immune responses of mice infected with vaccinia virus were studied, thus emphasizing the physiological importance of the effects of IL-2 during *in vivo* immune responses. However, missing from this report was an analysis of FOXP3 expression by the antigen-stimulated TCR-Tg CD4+ T cells.

References

1. Tran, Q., Glass, D., Uzel, G., Darnell, D., Spalding, C., Holland, S., and Shevach, E. (2009) Analysis of adhesion molecules, target cells, and the role of IL-2 in human FOXP3⁺ regulatory cell suppressor function. *J. Immunol.* **182**:2929–2938.

2. Pandiyan, P., Zheng, L., Ishihara, S., Reed, S., and Lenardo, M. (2007) CD4⁺CD25⁺FOXP3⁺ regulatory T cells induce cytokine deprivation-mediated apoptosis of effector CD4⁺ T cells. *Nat. Immunol.* **8**:1353–1362.

3. Pandiyan, P., and Lenardo, M. (2008) The control of CD4⁺CD25⁺FOXP3⁺ regulatory T cell survival. *Biol. Direct* **3**:1745–1757.

4. Kundig, T.M., Schorle, H., Bachmann, M.F., Hengartner, H., Zinkernagel, R.M., and Horak, I. (1993) Immune responses in interleukin-2-deficient mice. *Science* **262**:1059–1061.

5. Lyons, A. and Parish, C. (1994) Determination of lymphocyte division by flow cytometry. *J. Immunol. Meth.* **171**:131–137.

6. Fazekas de St. Groth, B., Smith, A., Koh, W.-P., Giris, L., Cook, M., and Bertolino, P. (1999) Carboxyfluorescein diacetate succinidyl ester and the virgin lymphocyte: a marriage made in heaven. *Immunol. Cell Biol.* **77**:530–538.

7. Fazekas de St. Groth, B., Smith, A., and Higgins, C. (2004) T cell activation: *in vivo* veritas. *Immunol. Cell Biol.* **82**:260–268.

8. Schorle, H., Holtschke, T., Hunig, T., Schimpl, A., and Horak, I. (1991) Development and function of T cells in mice rendered interleukin-2 deficient by gene targeting. *Nature* **352**:621–624.

9. D'Souza, W.N., and Lefrancois, L. (2003) IL-2 is not required for the initiation of CD8 T cell cycling but sustains expansion *J. Immunol.* **171**:5727–5735.

10. Zhang, X., Sun, S., Hwang, I., Tough, D.F., and Sprent, J. (1998) Potent and selective stimulation of memory-phenotype CD8⁺ T cells *in vivo* by IL-15. *Immunity* **8**:591–599.

11. Tsunobuchi, H., Nishimura, H., Goshima, F., Daikoku, T., Nishiyama, Y., and Yoshikai, Y. (2000) Memory-type CD8⁺ T cells protect IL-2 receptor alpha-deficient mice from systemic infection with herpes simplex virus type 2. *J. Immunol.* **165**:4552–4560.

12. Perez, O., and Nolan, G. (2002) Simultaneous measurement of multiple kinase states using polychromatic flow cytometry. *Nature Biotech.* **20**:155–162.

13. Krutzik, P., Clutter, M., and Nolan, G. (2005) Coordinate analysis of murine immune cell surface markers and intercellular phosphoproteins by flow cytometery. *J. Immunol.* **175**:2357–2365.

14. Long, M., and Adler, A. (2006) Cutting edge: paracrine but not autocrine, IL-2 signaling is sustained during early antiviral CD4 T cell response. *J. Immunol.* **177**:4247–4261.

15. O'Gorman, W., Dooms, H., Thorne, S., Kuswanto, W., Simonds, E., Kruzik, P., Nolan, G., and Abbas, A. (2009) The initial phase of an immune response functions to activate regulatory T cells. *J. Immunol.* **183**:332–339.

16. Gillis, S., Ferm, M.M., Ou, W., and Smith, K.A. (1978) T cell growth factor: parameters of production and a quantitative microassay for activity. *J. Immunol.* **120**:2027–2032.

17. Sojka, D.K., Bruniquel, D., Schwartz, R.H., and Singh, N.J. (2004) IL-2 secretion by CD4+ T cells *in vivo* is rapid, transient, and influenced by TCR-specific competition. *J. Immunol.* **172**:6136–6143.

18. Almeida, A., Zaragoza, R., and Freitas, A. (2006) Indexation as a novel mechanism of lymphocyte homeostasis: the number of CD4+CD25+ regulatory T cells is indexed to the number of IL-2-producing cells. *J. Immunol.* **177**:192–200.

19. Yu, A., Zhu, L., Altman, N., and Malek, T. (2009) A low interleukin-2 receptor signaling threshold supports the development and homeostasis of T regulatory cells. *Immunity* **30**:204–217.

20. Johnston, J., Kawamura, M., Kirken, R., Chen, Y.-Q., Blake, T., Shibuya, K., Ortaldo, J., McVicar, D., and O'Shea, J. (1994) Phosphorylation and activation of the JAK-3 Janus kinase in response to interleukin-2. *Nature* **370**:151–153.

Chapter 25

The Role of the IL-2r Chains in IL-2 Signaling, Consumption and Suppression of T Cell Proliferation

Twenty-five years ago, when we first could ask the question whether it makes any difference if a cell expresses just a few IL-2Rs or many IL-2Rs, only the IL-2Rα chain was known. Of course, as detailed earlier, the answer was an unequivocal "yes!" It does indeed make a difference.[1] Because cells differed in the density of IL-2Rα chains as much as three orders of magnitude, some cells could have only 100 sites/cell, while others could have as many as 100,000 sites/cell, with most cells distributed about a mean of ~1–10,000 sites/cell. The difference that we could uncover at the time, using $G_{0/1}$ synchronized cells, was that the IL-2Rα chain density determined how rapidly a cell progressed through G_1, passed the R-Point, and began to replicate DNA. The cells with the greatest density of IL-2Rα chains were the fastest. We concluded that the cell somehow counts the absolute number of triggered IL-2Rs over time, and only passes the R-Point when a critical number is surpassed.

Now a lot more is known about how IL-2 actually triggers the IL-2R, and how this occurs at the molecular level. However, one of the mysteries identified early on was the discrepancy in the density of the IL-2Rα chain versus the IL-2Rβ/γ chains, in that there was a mean of 100-fold more IL-2Rα chains (Fig. 25.1).[2,3] By comparison, there was always an equal but smaller number of IL-2Rβ/γ chains, which actually convey the proliferative signals.[4]

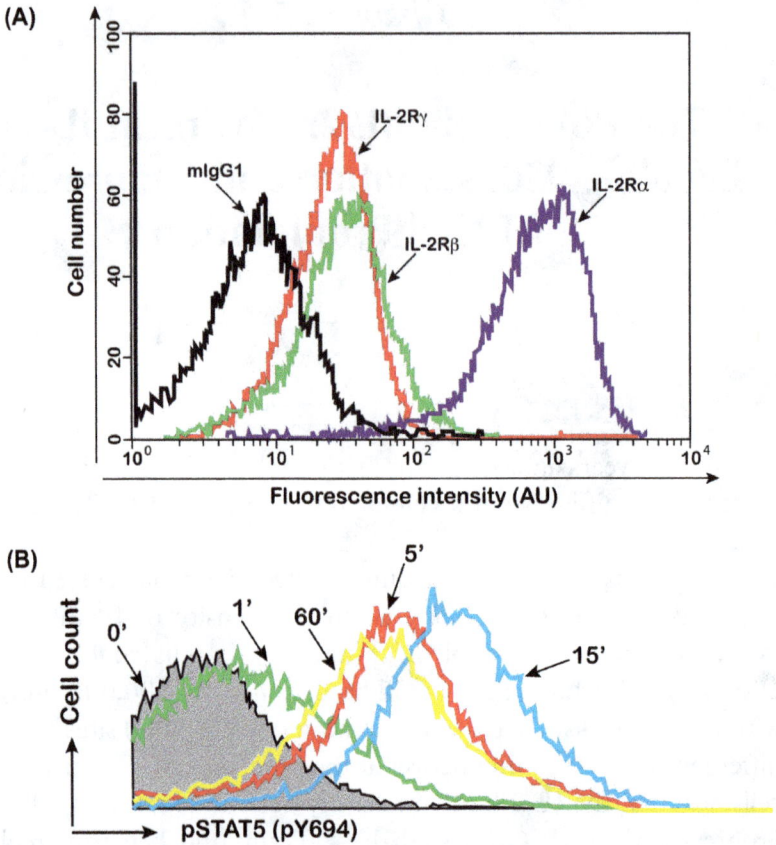

Figure 25.1:   A. The imbalance in expression of the IL-2Rα chains vs. IL-2Rβ/γ chains, and B. The time-course and log-normal distribution of IL-2-stimulated pSTAT5 expressions. A. Normal human T cells activated via anti-CD3 for 48 hours, followed by monitoring the expression of the IL-2R chains via flow cytometry. **B.** 48-hour activated IL-2R$^+$ human T cells were stripped of any bound IL-2 by a 5″ exposure to pH = 5.0, then washed and monitored at the depicted time intervals for pSTAT5 expression via flow cytometry. (From: Popmihajlov, Z. and Smith, K.A. Unpublished.)

Another advance that has occurred since the original experiments has been the accumulation of a detailed understanding of the signaling molecules activated via the triggered IL-2Rs, as well as the important genes expressed that move the cell through G_1 and into the S-phase. In addition, as detailed in the last few chapters, we have uncovered negative feedback loops that serve to dampen the response by inhibiting both IL-2 gene expression and IL-2 activity. Moreover, we now have the capacity to quantify many of these molecules at the single cell level via flow cytometry, and to relate the cellular activation status with the number of IL-2Rs triggered at the cell surface.

As already detailed, Gregoire Altan-Bonnet's group has pioneered a theoretical analysis of receptor triggering combined with single cell analysis of the activation of key molecules in known signaling pathways. Their initial work concerned how the quantal, digital activation of ppERK occurs upon TCR signaling.[5,6] Their approach using a combination of theoretical analysis and careful single cell experiments combined with quantification of the critical molecules has now been extended to IL-2 signaling.[7] The response to IL-2 was quantified initially by monitoring the generation of phosphorylated STAT5 (pSTAT5) following exposure of 150,000 TCR-activated cells to varying concentrations of IL-2 in 5 mL of medium (i.e. 30 cells/μL) for 10 minutes. Prior to proceeding, they ensured that the results from this short time interval were not confounded by endocytosis of the IL-2/IL-2R complexes. The generation of pSTAT5, along with the expression levels of the IL-2Rα chains and the IL-2Rβ chains were quantified by flow cytometry, and the amplitude of pSTAT5 and that of the IL-2 EC_{50} were both plotted against the IL-2R densities. The amplitude of pSTAT5 was monitored via flow cytometry after exposure to saturating IL-2 concentrations, while the IL-2 EC_{50} was determined by varying the IL-2 concentrations, and plotting those that yielded 50% of the maximum pSTAT5 amplitude. Plotting IL-2R subunit density against both the amplitude of pSTAT5 and the IL-2 EC50 parametrized the data. As shown in Fig. 25.2, the pSTAT5 amplitude increased as the density of both IL-2Rα (x-axis), and the IL-2Rβ (y-axis) increased. By comparison, the IL-2 EC_{50} became lower (i.e. the IL-2 binding affinity became higher), the greater the density of the IL-2Rα chain. Thus, at low IL-2Rα chain densities of only 100/cell, the IL-2 EC_{50} = 100 pM, while at high IL-2Rα chain densities of >10^5/cell, the IL-2 EC_{50} = 100 fM, with little influence from the IL-2Rβ chain density. This variation of 1000-fold in the IL-2 EC_{50} was entirely unexpected.

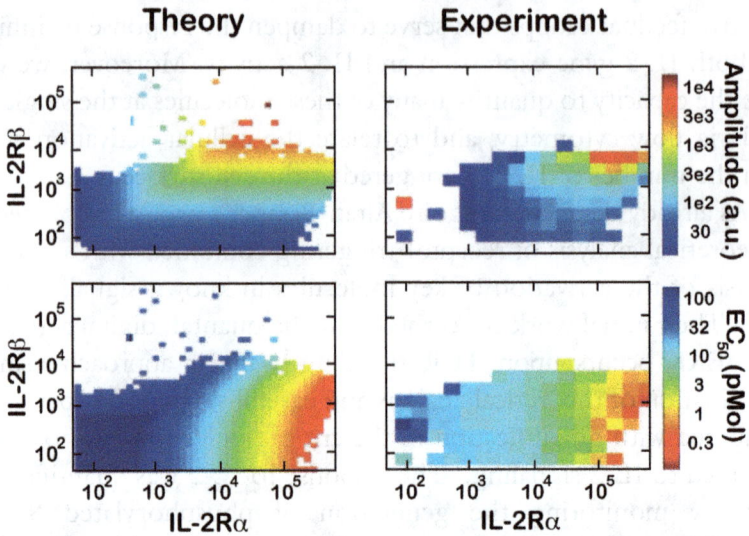

Figure 25.2:   IL-2 signaling of pSTAT5 depends upon IL-2R chain density. IL-2R⁺ T cells were exposed to varying IL-2 concentrations for 10 minutes followed by quantification via flow cytometry of intracellular pSTAT5 and cell surface density of IL-2Rα (*x*-axis) and IL-2Rβ chains (*y*-axis). Data are parametrized according to amplitude of pSTAT5 (top panels) or the IL-2 EC_{50} (bottom panels) as described in the text. Panels on left are from *in silico* modeling (Theory), and panels on right are derived from experiments (Experiment). (From: Feinerman, O. *et al.* 2010. In preparation.)

On the other hand, it was expected that the amplitude of pSTAT5 should be primarily determined by the densities of both of the IL-2R subunits, since the signals are generated by the trimeric IL-2Rα/β/γ complex. A high density of IL-2Rα chains could very well facilitate complexing of IL-2Rα chains with a much lower density of IL-2Rβ and γ chains via the law of mass action, as we originally conjectured.[8] However, it was unexpected that the *efficiency* or *affinity* of IL-2 binding would increase 1000-fold as a consequence a 1000-fold increase in IL-2Rα chain density; especially as the IL-2Rα chain itself does not contribute to signaling. In retrospect, Forsten and Laufenberger had modeled several different possible mechanisms to account for the wide discrepancy between the densities of IL-2Rα chains versus IL-2Rβ/γ chains, and found that the most plausible mechanism leading to densities of IL-2Rα chains above 10^5 copies/cell was that the IL-2Rα chains could enforce a ligand recapture phenomenon, which would establish an autocrine loop.[9] This would effectively change the search of the ligand for a binding site from a three-dimensional search in space to a two-dimensional search in the plane of the membrane. Thus, these new data now provide the first experimental evidence that this molecular mechanism actually operates.

Once experimental data were obtained, an *in silico* model was constructed, using the experimentally determined association and dissociation rate constants for the IL-2/IL-2Rα interaction ($\kappa = 10^7$ M^{-1}sec^{-1}; $\kappa' = 0.4$ sec^{-1}), which takes place in three-dimensional extracellular space, and describes the rapid on/off kinetics of low-affinity binding between IL-2 and the IL-2Rα. The second interaction occurs on the surface of each individual cell and involves the binding of the IL-2/IL-2Rα complex with IL-2Rβ to form a more stable trimeric molecular signaling complex with a much slower dissociation rate ($\kappa = 3.3^{-4}$ M^{-1}sec^{-1}; $\kappa' = 2.3^{-4}$ sec^{-1}). A more sophisticated three-step model where the IL-2/IL-2Rα complex binds first to IL-2Rβ, followed by the binding of the IL-2Rγ chain to the trimeric complex, did not change the parameters of the two-step model. It is noteworthy that the results from the *in silico* model mirrored the results found experimentally, while a classical model for IL-2 binding to a preformed IL-2R$\alpha/\beta/\gamma$ heterotrimeric complex with a constant affinity of 10 pM did not.[7]

By definition, nTreg cells express relatively high constitutive levels of IL-2Rα, with a geometric mean of ~10,000 sites/cell, while a few cells express even 10-fold higher densities. Like antigen-activated T$_{eff}$ cells, Treg cells also express detectable, but much lower densities of <1000 IL-2Rβ&γ chains, which can signal in response to IL-2. Moreover, one rapid response to IL-2 signaling, now known to be mediated by pSTAT5,[10] is a marked up-regulation of IL-2Rα chain expression, approximately 20-fold, as noted in our earliest experiments (2). When monitored experimentally, extremely low IL-2 concentrations, i.e. 60 fM, rapidly induce a further six-fold increase in IL-2Rα chain expression on Treg cells. Accordingly, this IL-2-mediated IL-2Rα

chain up-regulation provides a feed-forward mechanism, such that IL-2 further enhances the Treg capacity to bind IL-2 at extremely low concentrations, thereby further activating pSTAT5, which even further up-regulates IL-2Rα chain expression. Also, since FOXP3 is an IL-2/pSTAT5-induced gene, this feed-forward mechanism further guarantees the inability of the Tregs to produce IL-2, thereby restricting IL-2 concentrations available to those produced by the T_{eff} cells.

Since the Treg cells are sensitive to very low IL-2 concentrations, which might readily be produced by T_{eff} cells in a co-culture, it seemed entirely plausible that the IL-2 produced by T_{eff} cells might serve to rapidly up-regulate Treg IL-2Rα chain expression, and thereby facilitate Treg IL-2 binding, internalization and degradation, resulting in cytokine deprivation apoptosis, as shown by Pandiyan and co-workers.[11,12] When tested directly by co-culturing Tregs and naive T cells from TCR transgenics, together with specific peptide to activate the transgenic T_{eff} cells to produce IL-2, there was a marked up-regulation of IL-2Rα chains on Treg cells within 27 hours. This augmentation was entirely inhibited by a MoAb reactive with the IL-2Rα chain. Therefore, Treg cells act as dominant-negative regulators, scavenging any IL-2 produced by cells within their vicinity.

When TCR transgenic T cells are activated with increasing peptide concentrations, between 100 pM and 100 nM, the density of IL-2Rα chains expressed increases proportionally, spanning two orders of magnitude. Since each of the cells of the population are transgenic for the TCR, all TCRs have identical affinities for the peptide antigen. Accordingly, one cannot account for the difference in response to different antigen concentrations as due to varying affinities of the cells within the population. Rather, by analogy to the IL-2/IL-2R data, varying densities of TCRs on the individual cells within the population must account for the varying responses as monitored by the expression of IL-2Rα chain densities. Moreover, at low peptide concentrations, when fewer cells within the population express IL-2Rα chains, IL-2 accounts for much of the IL-2Rα chain expression, while at high peptide concentrations, when most of the cells express IL-2Rα chains, there is no IL-2 effect. It is noteworthy, that the TCR density on the individual cells most probably translates directly to the amount of IL-2/cell produced. It is also noteworthy that at low peptide concentrations, supplementation with excess IL-2 is insufficient to increase IL-2Rα chain expression to the densities observed at high peptide concentrations, underscoring the dual regulation of IL-2Rα chain expression by both the TCR and the IL-2/IL-2R.

Given the effect of IL-2 predominantly on the expression of the IL-2Rα chain in response to low antigen concentrations, it appeared that any suppressive Treg cell effect on T_{eff} cells would be most prominent at low antigen concentrations. Thus, monitoring the more immediate expression of T_{eff} cell IL-2Rα chains, rather than the longer-term endpoint of T_{eff} cell DNA synthesis, or the even more removed endpoint of cell division, the effect of Tregs was readily demonstrable. However at high peptide concentrations, e.g. >100 nM, when there is maximal IL-2Rα chain expression as well as maximal IL-2 production, there was no discernible suppressive effect of Tregs. Of course, these findings are entirely consistent with those of previous investigators, who reported that supplementation with exogenous IL-2 overcomes any Treg suppression of T_{eff} cell proliferation.[13]

From all of these data, it appeared that the density of IL-2Rα chain expression by both Treg cells and T_{eff} cells could very well be a critical parameter, one that most investigators have failed to consider. Because the density of IL-2Rα chain expression on either cell type dictates the efficiency of IL-2 binding, and thus internalization and degradation, if Treg cells have a higher density of IL-2Rα chain expression than T_{eff} cells, it would make them better IL-2 scavengers, actually *professional IL-2 scavengers*, particularly when IL-2 concentrations are low. By comparison, if T_{eff} cells have a higher density of IL-2Rα chain expression, they could also sense very low IL-2 concentrations, and thus should be less sensitive to IL-2 scavenging by Treg cells.

When these parameters are modeled *in silico*, if the IL-2Rα chain density on T_{eff} cells is set higher than the IL-2Rα chain density on Treg cells, only a few of these T_{eff} cells can circumvent Treg-mediated suppression. By comparison, if the IL-2Rα chain density on T_{eff} cells is set low relative to the IL-2Rα chain density on Treg cells, then higher concentrations of IL-2 are necessary to activate the T_{eff} cells, and Treg cell IL-2 consumption becomes more relevant, and suppression of these T_{eff} cells more apparent.

Taking into consideration the effect of varying levels of the density of IL-2Rβ chains, and focusing on the measurement of IL-2 signaling of pSTAT5 by single cells, a more complex *in silico* model was built. This model also incorporated measurements of IL-2 depletion, as well as both negative and positive feedbacks for IL-2 production and IL-2R densities. Simulating varying peptide antigen concentrations, which would result in the variation of both IL-2 production and IL-2R densities, T cell activation was modeled over 60 hours of activation, with and without Treg cells. This more dynamic model predicted that Treg cells inflict a "double hit" on T_{eff} cells activated at low peptide antigen concentrations. By depleting IL-2, Tregs limit the IL-2 concentration available, and the lower IL-2 concentration available limits the IL-2-mediated up-regulation of IL-2Rα chain density. Consequently, T_{eff} cells activated at low peptide antigen concentrations cannot maintain pSTAT5 activation high enough and long enough to promote cell cycle progression to DNA synthesis and cytokinesis. By comparison, at high peptide concentrations the same model predicts that T_{eff} cells that express IL-2Rα chain densities higher than the Treg cells can respond to lower IL-2 concentrations and still sustain high enough levels of pSTAT5 to signal cell cycle progression, despite IL-2 consumption by Treg cells.

These predictions were next tested experimentally. By consuming IL-2, Treg cells lead to lower IL-2 concentrations, but they also reduce T_{eff} cell IL-2 sensitivity by preventing efficient IL-2-mediated up-regulation of IL-2Rα density. Consequently, reduced densities of IL-2Rα chains along with lower extracellular IL-2 concentrations result in a "double hit" on pSTAT5 levels of T_{eff} cells. Thus, this "double hit" mechanism explains quantitatively how Treg cells suppress T_{eff} cells activated at low antigen concentrations, while permitting T_{eff} cells activated at high antigen concentrations to thrive and proliferate, thereby discriminating between auto-antigens, which are present at low concentrations and foreign antigens, which are present at high concentrations.

These considerations thus define the major differences between self and non-self peptides. According to the mechanisms responsible for positive selection during thymocyte maturation, only those T cells with low affinity for self-pMHC will be selected to survive and populate the periphery.[14] The other major characteristic that allows the immune system to distinguish between self and non-self pMHC molecules is their concentration. As long as self-pMHC molecules remain at a low density on the surface of cells, those TCRs with low affinity for these self-pMHC molecules will remain non-reactive. According to the Quantal Theory, only when the T cell receives the requisite number of triggered TCRs will enough IL-2 be produced and a high enough density of IL-2Rs be expressed, to activate a proliferative clonal expansion so that autoimmunity becomes manifest macroscopically, i.e. at the systemic level. Moreover, as we have seen, the amounts of IL-2 produced and the density of IL-2Rs expressed must surpass the governorship of IL-2R$^+$FOXP3$^+$ cells, which serve to scavenge the IL-2 produced by T_{eff} and thereby to dampen their expression of IL-2Rα chains.

References

1. Cantrell, D.A., and Smith, K.A. (1984) The interleukin-2 T-cell system: a new cell growth model. *Science* 224:1312–1316.
2. Smith, K.A., and Cantrell, D.A. (1985) Interleukin 2 regulates its own receptors. *Proc. Natl. Acad. Sci. USA* 82:864–868.
3. Smith, K.A. (1988) Interleukin-2: inception, impact, and implications. *Science* 240:1169–1176.
4. Russell, S., Johnston, J., Noguchi, M., Kawamura, M., Witthuhn, B., Silvennoinen, O., Goldman, A., Schmalsteig, F., Ihle, J., O'Shea, J., *et al.* (1994) Interaction of IL2R beta and gamma-c chains with JAK1 and JAK3: implications for XSCID and XCID. *Science* 266:1042–1045.
5. Altan-Bonnet, G., and Germain, R.N. (2005) Modeling T cell antigen discrimination based on feedback control of digital ERK responses. *PLoS Biology* 3:e356.

6. Feinerman, O., Veiga, J., Dorfman, J., Germain, R., and Altan-Bonnet, G. (2008) Variability and robustness in T cell activation from regulated heterogeneity in protein levels. *Science* **321**:1081–1084.

7. Feinerman, O., Smith, K., and Altan-Bonnet, G. (2009) Differential suppression of effector T cells by regulatory T cells derives from a highly dynamic IL-2 tug-of-war. *Submitted.*

8. Smith, K.A. (1989) The interleukin 2 receptor. *Annual Rev. Cell. Biol.* **5**:397–425.

9. Forsten, K., and Lauffenburger, D. (1994) The role of low affinity interleukin-2 receptors in autocrine ligand binding: alternative mechanisms for enhanced binding effect. *Mol. Immunol.* **31**:739–751.

10. Nakajima, H., Liu, X.-W., Wynshaw-Boris, A., Rosenthal, L.A., Imada, K., Finbloom, L.H., Henninghausen, L., and Leonard, W.J. (1997) An indirect effect of Stat5a in IL2-induced proliferation: a critical role for Stat5a in IL2-mediated IL2 receptor alpha chain induction. *Immunity* **7**:691–701.

11. Pandiyan, P., Zheng, L., Ishihara, S., Reed, S., and Lenardo, M. (2007) CD4+CD25+FOXP3+ regulatory T cells induce cytokine deprivation-mediated apoptosis of effector CD4+ T cells. *Nat. Immunol.* **8**:1353–1362.

12. Pandiyan, P., and Lenardo, M. (2008) The control of CD4+CD25+FOXP3+ regulatory T cell survival. *Biol. Direct* **3**:1745–1757.

13. Thornton, A.M., and Shevach, E.M. (1998) CD4+CD25+ immunoregulatory T cells suppress polyclonal T cell activation *in vitro* by inhibiting interleukin 2 production. *J. Exp. Med.* **188**:287–296.

14. Kisielow, P., Teh, H.S., Bluthmann, H., and von Boehmer, H. (1988) Positive selection of antigen-specific T cells in thymus by restricting MHC molecules. *Nature* **335**:730–733.

Chapter 26

T Cell Tissue-specific Autoimmunity

In 1972 Sir MacFarlane Burnet proposed that autoimmunity results from a genetic predisposition due to germline mutations that are propagated for the most part as silent heterozygous mutants, combined with random somatic mutations in relevant genes.[1] Of course, at that time the nature of the important molecules responsible for the generation of an immune response had yet to be discovered. Moreover, molecular genetics had yet to come into being. Therefore, exactly which genes might be responsible for this genetic etiology of autoimmunity remained entirely obscure and was unapproachable.

The understanding that there are negative feedback controls that operate to dampen immune responses, especially T cell immune responses, and the identification of the molecules that function in these negative feedback pathways, now provides the insight that Burnet lacked. Now finally we rest on firm evidence that disruption of any of the molecules in these negative feedback loops results in severe autoimmunity, in that the results of the gene deletion experiments are unambiguous. Deletion of IL-2, IL-10, TGFβ, the IL-2Rα chain, the IL-2Rβ chain, CTLA-4, PD-1, and FOXP3 all lead to similar, severe, polyclonal T cell proliferative multi-organ, rapidly lethal autoimmune diseases.

Even so, severe multi-organ polyclonal T cell autoimmune diseases are uncommon, with the notable exception of the IPEX syndrome. In this regard, it is noteworthy that ~1/3 of patients with the IPEX syndrome do not have mutations in the FOXP3 gene.[2] A recent report of the IL-2Rα chain deficiency responsible for an IPEX-like syndrome is instructive and illustrative of how a deficiency of the IL-2Rα chain leads to the paradoxical co-existence of immun-odeficiency and the polyclonal multi-organ lymphoproliferative autoimmune IPEX syndrome.[3] The patient inherited two mutated IL-2Rα chain alleles, and lacked entirely IL-2Rα chain expression. The IPEX syndrome became manifest in infancy at six weeks of age with the onset of severe diarrhea, insulin-dependent diabetes mellitus (IDDM), and the acute respiratory distress syndrome (ARDS), secondary to cytomegalic virus (CMV) infection. By two years of age, severe eczema, generalized lymphadenopathy, and hepato-splenomegaly developed, and by three years of age, hypothyroidism and autoimmune hemolytic anemia appeared. The normal expression of FOXP3 by the patient's CD4$^+$ T cells is shown in Fig. 26.1. Accordingly, given the normal expression of FOXP3, one cannot ascribe the patient's IPEX syndrome to a deficiency of FOXP3-mediated "active suppression." Rather, the syndrome is entirely consistent with dysregulated IL-2 binding and signaling.

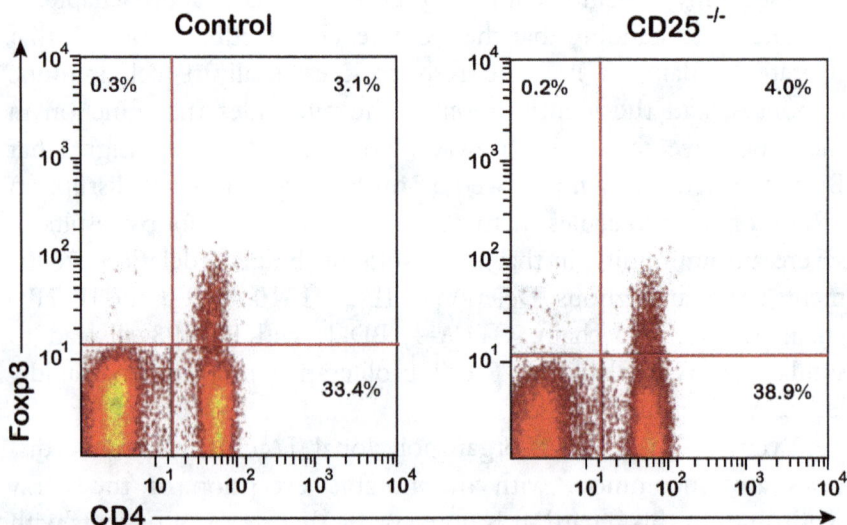

Figure 26.1:   **The normal expression of FOXP3 by CD4$^+$ T cells from a subject with homozygous IL-2Rα chain (CD25) deficiency.** PBMCs were isolated and reacted with MoAbs reactive with CD4 and FOXP3, without *in vitro* culture. (From: Caudy, A.A. *et al.* 2007. *J. Allergy Clin. Immunol.* **119**:482–487.)

The severe IPEX-like syndrome suffered by this patient contrasts with another patient with a rare homozygous missence mutation of the STAT5b gene.[4] This mutation primarily affected somatic growth, due to the lack of growth hormone effects, while the immune system remained relatively normal, presumably as a consequence of normal STAT5a expression. Repeated immunological evaluation throughout 20 years of life revealed only modest but consistently reduced circulating concentrations of both CD4[+] and CD8[+] T cells, as well as NK cells, but normal to elevated B cell concentrations. Moreover, it is important to note that this patient did not suffer from an IPEX-like syndrome, despite the fact that neither freshly isolated T cells nor T cells activated *in vitro* with PHA + IL-2 expressed detectable FOXP3. Instead, there was a selective defect in IL-2-promoted gene expression, with normal IL-2-induced expression of the IL-2Rγ chain and IFNγ, but IL-2Rα chain expression was only 20% of control levels and FOXP3 expression was absent. The mutant CD4[+] T cells were not anergic so that they could proliferate after allogeneic stimulation, consistent with the absence of FOXP3 expression by CD4[+]CD25[+] mutant T cells, and consistent with an intrinsic T cell abnormality, and the lack of FOXP3 suppression of IL-2 and the expression of other cytokine genes.

More commonly observed is T cell-mediated organ-, tissue-, and cell-specific autoimmune inflammation, such as is observed in type I diabetes mellitus (T1DM), Hashimoto's thyroiditis, multiple sclerosis, inflammatory bowel disease (IBD), and rheumatoid arthritis. Therefore, rather than a polyclonal T cell abnormality affecting multiple tissues/organs, it is necessary to postulate a mechanism(s) that could explain antigen-specific T cell-mediated inflammatory immunopathology of individual cells, tissues, or organs.

Burnet introduced the concept of the "forbidden clone" as a variation of his Clonal Selection Theory, to account for tissue-specific autoimmunity.[1] In this instance, he envisioned mutations that allowed an auto-reactive "forbidden clone" to escape deletion during development. He also introduced the notion that "sequential mutations" might be required to develop full-blown autoimmune disease. In particular he hypothesized that in addition to persistence of the forbidden clone, sequential mutations that permit *unrestricted clonal proliferation* might be necessary.

According to Burnet:

> "The chance of any given somatic mutation occurring in a given cell is always very rare. If its frequency is expressed as $1/n$, n will probably be of the order 10^5–10^8. The probability of the same cell undergoing two unrelated mutations will be $1/n' \times 1/n''$, which will usually be prohibitively small. There are, however, two ways that sequential mutations can occur. If mutation results in a cell with such a *proliferative advantage* that in a reasonable period of time it has, say, 10^6 descendents all carrying the mutant characteristic, this is a population of cells in which mutation #2 can occur."

In this instance, Burnet envisions that the first mutation provides the cell with a proliferative advantage, so that there will be a greater probability of a second mutation, which would confer on the cell its capacity to react with the auto-antigen, although given the ambiguity of triggering either T cells or B cells at the time, it is difficult to envision what genes Burnet had in mind.

Burnet proposed that the development of autoimmunity is analogous to the development of cancer, which is time-dependent, and consistent with the sequential mutation hypothesis. Thus, he pointed out the oxymoronic "stochastic regularity" in the age incidence of cancer, which yields a straight increasing line when the age-specific mortality per 10^5 individuals is plotted on a $\log_{10}$ scale versus a $\log_{10}$ scale of age, as shown in Fig. 26.2. According to Burnet, such regularity interested statisticians and epidemiologists, and led to the postulate that the most likely explanation is that the increasing straight line relationship between incidence and age is a rare somatic mutation, which may occur at any time during life, and which confers a proliferative advantage on a single cell compared with similar but un-mutated cells. Subsequently, one of the progeny of such a cell then suffers another stochastic mutation, which further increases the proliferative advantage until an eventual clone of cells is recognized macroscopically as a malignancy.

Burnet suggested that "a forbidden clone" of autoimmune cells has a close formal resemblance to a conditioned neoplasm, which led

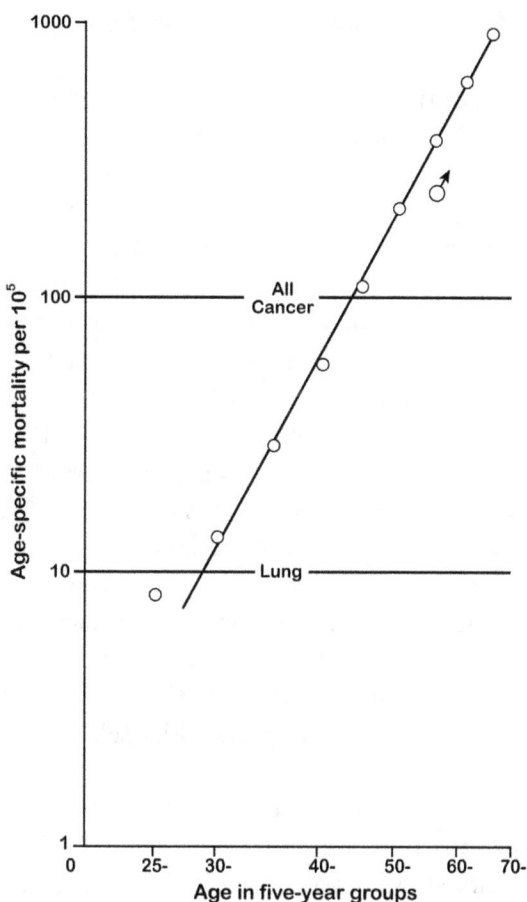

Figure 26.2: Stochastic regularities in the age incidence of cancer. Age-specific incidence of all cancers in males of cohort 1886–1890, England and Wales. Age-specific mortality per 10^5 plotted against age in years, both scales being logarithmic. (Redrawn from: *Autoimmunity and Autoimmune Disease. A Survey for Physician or Biologist*. 1972. Macfarlane Burnet. F.A. Davis Company. Philadelphia. Fig. 4, p. 21.)

to an analysis by stochastic mathematical methods of the age-specific incidence curves of a wide range of autoimmune diseases. The results are summarized as follows:

1) When predisposition to the disease is limited to one or more groups of people within the general population, the specificity of

such groups is determined by genetic factors (a.k.a. germline mutations).
2) The initiation of the disease process is contingent upon the occurrence of a random somatic mutation.
3) An interval elapses between the last random somatic mutation initiating the event and the age of onset of symptoms and signs of disease, which can be quite long, i.e. months, years, or even decades.

Of course, missing from these analyses and hypotheses made almost 40 years ago was an understanding of how the cells proliferate and how molecules of the immune system function. Now it is known that immature thymocytes are only positively selected if their TCRs react with low affinity to self-pMHC complexes so that they can fully mature and exit the thymus to populate the periphery.[5] Consequently, it follows that each and every mature peripheral T cell has the inherent capacity to react with at least one self-pMHC complex. The fact that they usually do not react can be explained by the fact that the selection process favors those self-pMHC complexes that react with low affinity, but also that are present in low concentrations in the periphery. Thus, according to the Quantal Theory, potential self-reactivity is held in check by ensuring that the number of triggered TCRs remains well below the number necessary to result in productive activation of the T cell and the transcriptional activation of the IL-2 gene, which is responsible for clonal expansion, and other cytokine genes, which are responsible for their differentiation to become T_{eff} cells.

Accordingly, it is now possible to see how tissue or cell-specific autoimmune reactivity might originate. T cell-mediated reactivity to a self-pMHC antigen might occur should the density of self-pMHC on a given tissue cell become high enough to trigger even low affinity TCRs. In normal circumstances this possibility remains remote, although one can imagine that environmental issues, such as inflammatory tissue damage due to infection or sterile inflammation might suffice to increase the density of the self-pMHC beyond the threshold that could lead to productive activation of an individual T cell. However, this in and of itself should not be enough to lead to the persistence and expansion of the selected T cell clone, given the quorum sensing effects of a population of cells with a significant population of anergic IL-2R$^+$FOXP3$^+$ IL-2 consuming T cells.

Another situation that could lead to T cell reactivity to a self-pMHC epitope might be a predisposition of an altered, i.e. lower, quantal number of triggered TCRs necessary for activation. This situation can be envisioned, should a germline mutation in any of the genes encoding the molecules involved in the negative feedback loops that govern the number of triggered TCRs necessary for activation. Moreover, given a gene dosage effect, a heterozygous mutation in a single gene may be insufficient to generate actual disease, unless paired with a random mutation in a second critical gene encoding the same or a separate molecule in the negative feedback loop.

Thus, according to the Quantal Theory, only two general situations can lead to the reaction to self-pMHC epitopes, either an increase in the density of self-pMHCs molecules, or a decrease in the number of triggered TCRs necessary for activation. In either instance, at least one hallmark would be, as Burnet predicted, a forbidden clone would be responsible for the autoimmune destruction of a specific cell, tissue, or organ.

References

1. Burnet, F. (1972) *Auto-immunity and auto-immune disease; a survey for physician or biologist.* F.A. Davis Co. Philadelphia, USA. p. 243.
2. Gambineri, E., Torgerson, T., and Ochs, H. (2003) Immune dysregulation, polyendocrinopathy, enteropathy, and X-linked inheritance (IPEX), a syndrome of systemic autoimmunity caused by mutations of FOXP3, a critical regulator of T cell homeostasis. *Curr. Opin. Rheum.* 15:430–435.
3. Caudy, A., Reddy, S., Chatila, T., Atkinson, J., and Verbsky, J. (2007) CD25 deficiency causes an immune dysregulation, polyendocrinopathy, enteropathy, X-linked-like syndrome, and defective IL-10 expression from CD4 lymphocytes. *J. Allergy Clin. Immunol.* 119:482–487.
4. Cohen, A., Nadeau, K., Tu, W., Hwa, V., Dionis, K., Bezrodnick, L., Teper, A., Gaillard, M., Heinrich, J., Krensky, A. *et al.* (2006) Decreased accumulation and regulatory function of CD4$^+$CD25 high T cells in human STAT5b deficiency. *J. Immunol.* 177:2770–2774.
5. Kisielow, P., Teh, H.S., Bluthmann, H., and von Boehmer, H. (1988) Positive selection of antigen-specific T cells in thymus by restricting MHC molecules. *Nature* 335:730–733.

Chapter 27

Type 1 Diabetes Mellitus (T1DM), a Prototypic Genetic Autoimmune Disease with a Tie to IL-2

Given Burnet's hypothesis that a "forbidden clone" of auto-reactive lymphocytes could be responsible for autoimmune destruction of specific cells, tissues, and organs, what evidence has accumulated to date that supports such a hypothesis? The most data accumulated in this regard are found in studies of T1DM.

It is now dogma that T1DM is caused by T cell-mediated immunological destruction of the insulin-producing pancreatic beta (β) cells. Studies in both humans and the non-obese diabetic (NOD) mouse autoimmune model of T1DM have suggested that the classical model of genetic susceptibility to disease involves a relatively small number of genes that contribute large effects to disease pathogenesis, together with a much larger number of genes that make smaller contributions. Recent highly powered genome-wide association (GWA) studies comparing a large number of individuals with T1DM with disease-free control subjects have enabled the discovery of a number of genetic elements *associated* with T1DM. Fifteen chromosomal regions harboring several candidate T1DM-associated gene loci have now been identified. What is perhaps most telling about the genetic elements identified thus far, is the fact that many harbor genes that are known to function either as antigen recognition molecules, or negative regulators of the immune response.

With regard to a T cell clonal derivation of T1DM, the first chromosomal region found to be associated with susceptibility to

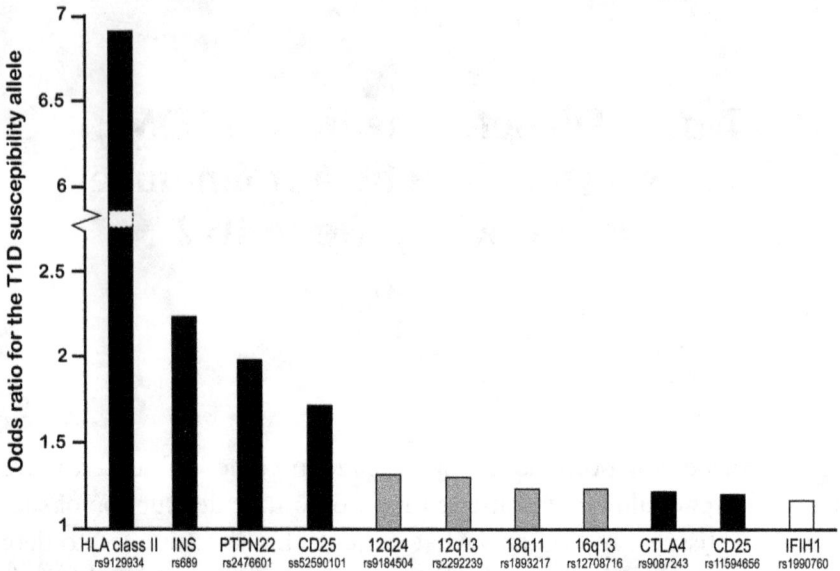

Figure 27.1: Odds ratios for the susceptibility allele for the 10 independent Ti1DM-associated genes or regions. The filled black bars indicate previously known associated genes and regions. The open bar indicates the *IFIH1* region identified by the nsSNP genome scan, and the filled grey bars were identified by the WTCCC Affymetrix 500K scan, and confirmed by the studies in this report. (Redrawn from: Todd, J.A. *et al.* 2007. *Nature Gen.* **39**:857–864.)

T1DM is the Class II MHC gene region on chromosome 6p21, discovered 20 years ago. It also yields the highest relative risk for developing T1DM among all known T1DM-associated gene loci. The odds ratios for the susceptibility allele for 10 of the independent T1DM-associated genes or regions are shown in Fig. 27.1.[1] These data imply that some HLA molecules must preferentially bind self-peptides derived from the pancreatic β-cells. In support of this conjecture, genes encoding HLA-DQ and –DR and their murine counterparts (I-A and I-E respectively) are thought to harbor the etiological polymorphisms accounting for most of the HLA Class II-T1DM associations. Susceptibility alleles at these loci share certain structural similarities, particularly at pockets in their peptide-binding grooves, thereby accommodating specific peptide residues, implicating that they can bind specific T1DM-associated peptides and thus select specific TCRs.

As a β-cell specific source of peptides that might bind to specific HLA molecules, another gene locus found to be associated with human T1DM maps to the insulin-coding gene. The susceptibility to T1DM is associated with reduced transcription of insulin mRNA in the thymus. Thus, if insulin auto-reactive T cells are not appropriately deleted in the thymus during development, it is easy to see how given the appropriate HLA Class II alleles, insulin-reactive CD4$^+$ T cell clones could be selected, activated and expanded.

Beyond HLA and potential self-reactive peptides, among other gene loci associated with T1DM that stand out include a phosphatase negative regulator of TCR signaling, protein tyrosine phosphatase non-receptor type 2 (*PTPN2*) and *CTLA-4*. Although these still remain simply associations and no cause/effect relationships have been established, these links at least point the way to search for more associations with negative regulators, and to investigate how the association with these gene loci could possibly lead to loss of self–non-self discrimination.

One additional gene locus of note associated with T1DM is a locus that harbors the genes encoding the IL-2Rα chain, as well as the IL-15Rα chain. Since both of these genes are so involved with T cell activation and reactivity, it is difficult not to wonder whether defects in the IL-2 negative regulation of T cells contributes a great deal to the loss of self–non-self recognition. In support of this conjecture, the IL-2Rα chain gene has also been found to be associated with multiple sclerosis (MS), as well as celiac disease.

Even so, all of these genetic studies thus far have only given us a clue as to gene abnormalities that could contribute a familial (germ line) genetic predisposition to autoimmune destruction of the insulin-producing pancreatic β-cells. For the most part, these gene associations have not led to readily testable hypotheses that could suggest cause/effect, such that one could envision an intervention that could reverse the disease process once identified, or even more important, prevent the development of the disease process. The one exception is compelling evidence that has been generated recently by experiment are data examining a gene locus spanning 780 kb termed insulin-dependent diabetes 3 (*idd3*), on mouse chromosome 3.[2] This locus has been narrowed to ~650 kb and contains five known genes, including *Tenr*, *Il2*, *Il21*, *Cetn4*, *and Fgf2*, as well as two predicted genes of unknown function.

In a series of detailed experiments, it was established that *idd3*-dependent susceptibility to T1DM localizes to the *il2* gene, but there were no mutations within the coding region of the gene. Instead, susceptibility was found to be dependent on single nucleotide polymorphisms (SNPs) in the 10,000 bp upstream from the gene, and the phenotype resulted in a 50% decrease in TCR-activated IL-2 production. Additional experiments indicated that the *il2* susceptible genotype determines a difference in the *proportion* of cells capable of producing *il2* mRNA. Thus, the *idd3* B6WT genotype (i.e. resistant) supported the production of detectable intracellular IL-2 in a *higher fraction* of cells than the *idd3* NOD (i.e. susceptible) genotype. Moreover, analysis of *Il2* pre-mRNA revealed that the T1DM-susceptible NOD *Il2* allele is ~ half as active as the WT allele. It is also noteworthy that in contrast to the findings for *Il2*, no differences were observed in *Il2* transcription.

To prove that T1DM susceptibility is actually caused by a 50% decreased capacity to produce IL-2, one dose of a susceptible *Idd3* region carrying a targeted mutation of *Il2* was introduced into mice carrying a protective B6 *Idd3* region. As expected, activated splenic T cells from these mice produced ~50% less IL-2 and contained ~half as many IL-2+ T cells as isolated from controls. These experimental mice developed a significantly higher frequency of T1DM than controls, thereby leading to the conclusion that *Il2* is a major component of the *Idd3* susceptibility locus.

Exactly why only a ~two-fold decrease in IL-2 gene expression should lead to T1DM is another question. These investigators suggested that deficient production of IL-2 by CD8+ T cells leads to a diminution of CD4+CD25+FOXP3+ Tregs, which then predisposes to the activation of auto-reactive CD8+ T cells. They postulated that IL-2-activated T-Reg cells normally suppress dendritic cell maturation, which inhibits the cross presentation of β-cell auto-antigens to potential auto-reactive CD8+ T cells, thereby maintaining homeostatic control of auto-reactive T cells and preventing T1DM disease progression.

In another study, Christopher Goodnow's team approached directly how variation in IL-2 activity affects the control of islet-specific T cells.[3] To determine the role of IL-2 in CD4$^+$ T cell reactivity versus tolerance to pancreatic islet antigens, IL-2 (-/-) mice were crossed with mice transgenic for T cells that recognize peptide 41–61 from hen egg-white lysozyme (HEL), together with mice expressing the insHEL transgene, which is controlled by the rat insulin promoter and mirrors expression of insulin itself: high expression of HEL occurs in pancreatic islet β-cells, whereas HEL expression in thymic epithelial cells depends on *Aire*.

In TCR:insHEL double transgenic mice, thymic HEL expression deletes most of the high density TCR clonotype HEL-reactive CD4$^+$ T cells, leaving only low density TCR CD4$^+$ T cells to populate the periphery. These cells are unable to destroy the HEL-expressing pancreatic islet β-cells and precipitate diabetes, although they cause subclinical insulitis from an early age. However, when these mice are crossed to IL-2 (-/-) mice, the lack of IL-2 precipitates diabetes in 100% of animals. By comparison, less than 20% of double transgenics on a WT or IL-2 (+/-) background develop diabetes.

When examined directly, both positive and negative selection of IL-2-deficient HEL TCR clonotypic double transgenic cells were normal compared with WT mice. Therefore, the lack of any requirement for IL-2 in thymic negative selection provides strong evidence against a role for IL-2 in this major mechanism of organ-specific self-tolerance.

It is noteworthy that IL-2 deficiency had no effect on the relative numbers of CD4$^+$FOXP3$^+$ islet-reactive TCR$^+$ T cells in IL-2 deficient versus IL-2 sufficient mice. In marked contrast, IL-2 deficiency resulted in the peripheral expansion, a ~250% increase, of CD4$^+$FOXP3$^-$ with low densities of islet-reactive TCR. These FOXP3$^-$ cells were also CD25$^-$, suggesting that optimal FOXP3 and CD25 expression are both normally dependent on IL-2. The reduction in surface islet-specific TCR densities on these cells also was not due to antigen-induced activation down-regulation, as it was not observed in the HEL-specific T cell population in diabetic control mice. Under normal IL-2-sufficient conditions, this population of TCR low density islet-reactive T cells does not precipitate diabetes. However, in the absence of IL-2, the increased frequency of these TCR low density islet reactive T cells is associated with the development of diabetes in 100% of animals.

Accordingly, these experiments do not support the premise that a deficiency of $CD4^+FOXP3^+$ "regulatory" T cells results in the expansion of low density TCR islet-reactive $CD4^+FOXP3^-$ T cells that then cause diabetes. Instead, the data are consistent with an immune dysregulation as a consequence of IL-2 deficiency that leads to a paradoxical loss of self-discrimination. In this reductionist approach to examine autoimmunity, it is important to note that because TCR transgenics were used, all of the $CD4^+$ TCRs have the same affinity of reactivity with only one HEL peptide. Therefore, affinity of interaction between the TCR and pMHC complex cannot account for the results. In this respect, the only way that the TCR transgenic T cells could escape negative selection in the thymus is by means of a lowered TCR density. On a WT genetic background, these T cells are inefficient in causing autoimmune destruction of the islet β-cells, even when the antigenic HEL peptide is expressed at a high concentration. Evidently, the quantal number of pMHC-triggered TCRs is not surpassed, so that most T cells remain quiescent. However, when these double transgenic mice are crossed to IL-2 (-/-) mice, autoimmunity ensues in 100% of the animals.

The absence of IL-2 in a situation of a persistent antigen, together with persistence of low TCR density transgenic T cells, results in a complete lack of IL-2-mediated negative feedback regulation, not only by the loss of FOXP3, but the loss of CTLA-4 and IL-10 as well. Although Goodnow's group did not monitor for the expression of these IL-2-induced genes, future experiments can now be performed to monitor these parameters. If the transgenic $CD4^+$ T cells lack IL-2-induced FOXP3 expression, one would predict that there is no FOXP3 feedback suppression of cytokine gene expression, especially of the $IL-2R\gamma_c$ chain utilizing cytokines such as IL-4, IL-7, IL-9, IL-15, and IL-21, all of which could account for the peripheral expansion of low TCR density transgenic T cells. Moreover, the lack of IL-2-mediated CTLA-4 expression by these cells would allow their activation via pMHC without a CTLA-4-mediated check on TCR signaling as would be expected to occur normally. Finally, the lack of IL-2-induced expression of IL-10, would have the effect of the absence of an IL-10 anti-proliferative effect, usually mediated by this IL-2-promoted negative feedback loop.

The other aspect of this study that presumably is absent normally, relates to the use of the insHEL transgenic mice, which sets up a situation where the islet "auto-antigen" is expressed at high levels. As we have detailed, when the antigen concentration is low, the quantal activation of T cells, even those with high TCR densities and high affinities, is insufficient to produce T cell activation of *IL-2* and *IL-2Rα* chain gene expression. Thus, in order for loss of self versus non-self discrimination to occur, it may be necessary for auto-antigens to be expressed at abnormally high concentrations. If so, one might expect activation of a single "forbidden clone" as proposed by Burnet by a single pMHC complex.

In this regard, investigators have now developed T cell clones from NOD mice and identified some, but not all, of the peptides recognized (reviewed in Ref. 4). It is noteworthy that both CD4$^+$ and CD8$^+$ pMHC-specific clones have been identified thus far, and most investigators have found that both types of T cells are necessary to induce T1DM in NOD mice. Accordingly, Burnet's "forbidden clone" may well have been too simplistic, given the complexity of the cells and their interactions involved in mounting a productive immune response, whether to a non-self foreign antigen or to a self-antigen.

However, CD4$^+$ T cells are thought to be essential in both the early and late stages of the disease, as evidenced by the capacity of multiple CD4$^+$ T cell clones to transfer disease. In addition, anti-CD4 therapy can prevent the onset of disease in NOD mice. Over two dozen islet-specific auto-antigens have been implicated in the initiation and/or pathogenesis of T1DM. However, the antigenic specificities of the majority of diabetogenic T cell clones isolated from NOD mice still remain unknown. More important, a key question that remained unresolved until recently is the extent to which auto-antigenicity and pathogenicity are linked. In studies monitoring a panel of islet-specific TCRs, relatively few were found to promote islet entry and β-cell destruction, in the absence of other T cells and B cells, so that auto-reactivity is not equivalent to pathogenicity.[5] Auto-reactive TCRs can be segregated into three phenotypic groups:

1) TCRs that fail to mediate islet entry, but may play a role in later stages of the disease.
2) TCRs that mediate islet infiltration, but not β-cell destruction, and thus may only contribute significantly to T1DM if accompanied by diabetogenic T cells.
3) TCRs that mediate insulitis and β-cell destruction, and thus may be key diabetogenic TCRs.

Burnet's concept of the forbidden clone as responsible for tissue antigen-specific autoimmunity was based on the premise that during development all clones that react with self-antigens would be deleted. We now know that as a consequence of positive selection in the thymus, all cells that leave the thymus and populate the periphery have potential self-reactivity. Thus, the question becomes as to how these self-reactive clones are kept in check, or fail to react. The Quantal Theory predicts that peripheral tolerance is normally achieved through a combination of low self-antigen concentrations, low TCR affinity for the self-pMHC complex, several different negative feedback loops operative to dampen T cell activation, and finally an IL-2 quorum sensing effect mediated by the various cells comprising the total cell population. Accordingly, these parameters can now serve as a working hypothesis for future investigation of T cell-mediated tissue specific autoimmunity. Thus, autoimmunity might be due to an abnormally increased expression of self-peptides, together with mutations involving genes encoding molecules that participate in negative feedback control of T cell-mediated reactivity.

References

1. Todd, J., Walker, N., Cooper, J., Smyth, D., Downes, K., Plagnol, V., and *et al.* (2007) Robust associations of four new chromosome regions from genome-wide analyses of type 1 diabetes. *Nature Genetics* **39**:857–864.

2. Yamanouchi, J., Rainbow, D., Serra, P., Howlett, S., Hunter, K., Garner, V., Gonzales-Munoz, A., Clark, J., Veijola, R., Cubbon, R., *et al.* (2007) Interleukin-2 gene variation impairs regulatory T cell function and causes autoimmunity. *Nat. Gen.* **39**:329–337.
3. Liston, A., Siggs, O., and Goodnow, C. (2007) Tracing the action of IL-2 in tolerance to islet-specific antigen. *Immunol. Cell Biol.* **85**:338–342.
4. Haskins, K. (2005) Pathogenic T-cell clones in autoimmune diabetes: more lessons from the NOD mouse. *Adv. Immunol.* **87**:123–162.
5. Burton, A., Vincent, E., Arnold, P., Lennon, G., Smeltzer, M., *et al.* (2008) On the pathogenicity of autoantigen-specific T cell receptors. *Diabetes* **57**:1321–1334.

Chapter 28

The Pathogenesis of Leukemia —
Loss of Normal Quantal Growth Control

As a result of progress that has occurred over the past 50 years, we now have a reasonably complete understanding of the molecular pathogenesis of at least one type of cancer, i.e. chronic myelocytic leukemia (CML) (reviewed in Ref. 1). CML results from a mutation that circumvents the normal cytokine/receptor signaling of quantal cell cycle progression. The mutation, originally discovered by Peter Nowell in 1960 and termed the Philadelphia (*Ph*) chromosome,[2] results in the constitutive activation of a protein tyrosine kinase (PTK), the cellular proto-oncogene c-Abl,[3] which persistently phosphorylates one of the *src* family kinases, Hematopoietic cell Kinase (HcK), which in turn persistently phosphorylates and activates STAT5.[4] Persistently activated STAT5 then circumvents the normal hematopoietic cytokine/receptor quantitative control of the cell cycle, and results in the continuous expression of genes necessary for both cell cycle progression (e.g. the cyclin D genes), and anti-apoptotic genes important for cell survival (e.g. BclX).[5]

As described so presciently by Peter Nowell over 30 years ago, the clonal evolution of a tumor proceeds from a *founding* mutation that gives an individual cell a growth advantage over its normal counter-parts, so that its progeny survive and accumulate to form either a solid or a liquid tumor.[6] The BCR/ABL reciprocal translocation between chromosomes 9 and 22 serves this purpose,[7–9] and results in the abnormal accumulation of the progeny of a single hematopoietic stem cell (HSC). These cells are capable of maturing into polymorphonu-clear leukocytes, and elevated platelets and erythrocytes also occur.

Eventually, the autonomous growth of these cells and the accumulation of millions of progeny cells result in a much higher probability for additional mutations that render the cells with an additional selective growth advantage, which then increases the virulence of the malignancy.

Of utmost importance, the subsequent mutations usually do not occur for several years, so that there is a long lag-time from the first to the subsequent mutations. This time interval affords a window whereby treatment can be instituted to markedly reduce the tumor cell population before the more virulent malignant cells arise. For chronic phase CML, before the accelerated phase, a.k.a. blast crisis, the targeted drug Gleevec, which blocks the kinase activity of the BCR/ABL oncogene, is very effective, resulting in a complete hematological remission in >98% of individuals.[10] Gleevec essentially revolutionized oncology because it demonstrated that by inactivating a single molecule thought to cause the malignancy, it is possible to kill the mutated cells without major side effects on normal cells. However, if one waits until blast crisis ensues, the overall response rate is ~55%, with only 10% of patients achieving a complete remission.[11] Even so, this response rate in the more genetically complex blast crisis cells indicates that even these cells can still be dependent on a single *founding* mutation. In addition, these data indicate that if targeted therapy is available, patients should be treated early, before mutational progression to a more virulent malignancy.

Accordingly, CML represents the most completely understood example of a cancer that we have today. The question now is whether CML represents the exception or the rule. Many cancer researchers still ascribe to the notion that cancer only results as a consequence of several sequential mutations, as hypothesized by Burnet. However, as originally proposed by Nowell, the first or *founding* mutation, which is now called the *driving* mutation as a result of the CML/Gleevec experience, must by definition impart a growth advantage to the cell, and not simply block maturation, a former popular theory of malignancy. It is now evident that persistent activation of the cytokine/receptor/JAK-STAT/cyclinD pathway can result in autonomous cell cycle progression. There may well be mutations in a myriad of upstream genes that can feed into this final common pathway to ultimately usurp the normal quantal cytokine control of DNA replication and mitosis. However, there are primarily two members of the STAT family that are emerging as critical in the pathogenesis of malignancy, STAT3 and STAT5.[12-14] STAT3 is important for the many malignancies associated with chronic inflammation, as it is normally activated by the pro-inflammatory cytokine IL-6, which is produced in high amounts by macrophages. As the expression of IL-6 is itself activated by both sterile and non-sterile inflammation that trigger toll-like-receptors (TLRs) on phagocytic APCs, a chronic inflammatory environment is conducive to the persistent activation of STAT3.[15]

STAT5 is important in most of the hematological malignancies, because of its primary normal role to signal cell cycle progression in response to many of the interleukins and the hematopoietic cytokines. Thus, in addition to its role in CML, the persistent activation of STAT5 by activating point mutations in JAK2, which is normally regulated by erythropoietin (EPO), thrombopoietin (TPO) and granulocyte colony stimulating activity (G-CSA), results in most of the cases of the other chronic myeloproliferative diseases, including polcythemia vera, essential thrombocytosis, and myelofibrosis (reviewed in Ref. 16). In addition, many of the acute myeloid leukemias have also now been found to result from mutations in genes that regulate other PTKs, such as KIT and Flt3Rs, which also activate STAT5. Moreover, STAT5 is downstream of the prolactin-R as well as the EGF-R. Consequently there is mounting evidence that persistent activation of STAT5 may play a primary role in the malignant transformation of common epithelial cancers, e.g. breast, colon, prostate and lung cancers (reviewed in Ref. 17).

The cytokine/receptor/JAK/STAT/cyclinD pathways have emerged as so important in cancer pathogenesis because of their primary roles of normally supplying the molecular signals that promote cell cycle progression to move beyond the R-point, thereby initiating quantal DNA replication. As shown in Fig. 28.1, in T cells, antigen recognition initiates cell cycle competence, or G_0/G_1 transition, which in molecular terms is manifest by the expression of the IL-2

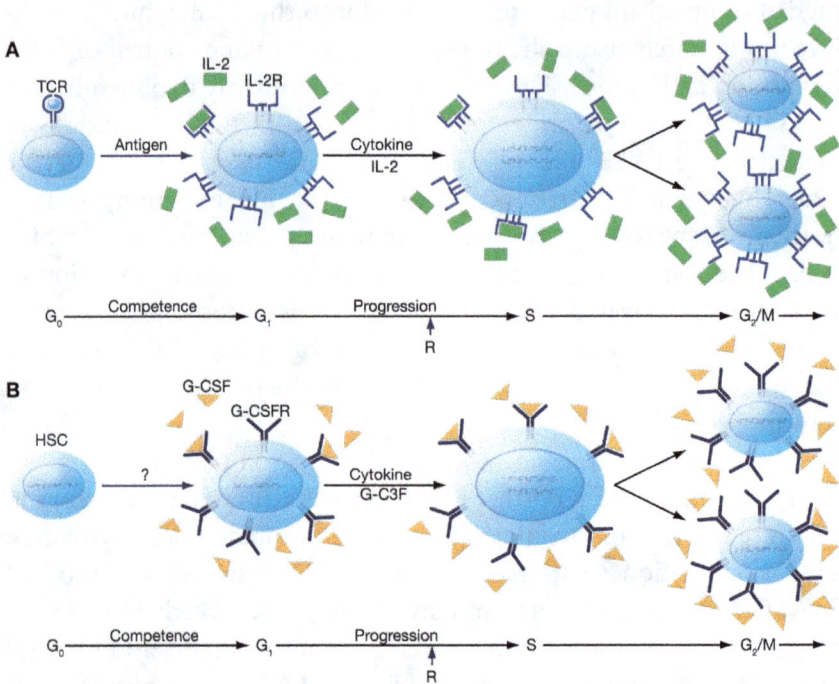

Figure 28.1:   Cell cycle competence and progression. A. Lymphocytes are normally in the resting, or G_0 phase of the cell cycle. Signals from the TCR promote the transition from G_0 to G_1, which renders the cells competent to progress further through the cell cycle by stimulating the expression of genes encoding IL-2 and the IL-2Rα chain. Subsequently, the IL-2/IL-2R interaction is responsible for blastic transformation and G_1 progression, which moves the cell through the G_1 phase of the cell cycle to the point that the cells no longer require signals from the IL-2/IL-2R interaction (i.e. past the R-point) to enter the S-phase and subsequently undergo mitosis. **B.** Hematopoietic Stem Cells (HSCs), like resting mature T lymphocytes, are also normally in G_0. Unlike the T cell lineage, little is known regarding the cytokines and receptors responsible for stimulating the HSCs to leave G_0 and enter the cell cycle. Even so, once this G_0-G_1 transition occurs, and an HSC has given rise to a committed granulocyte precursor, the cytokines responsible for G_1 progression are known, and this is exemplified in the figure by G-CSF acting on a committed granulocyte precursor to promote the progression from G_1 to the S, G_2, and M phases. (From: Smith, K.A. and Griffin, J.D. 2008. *J. Clin. Invest.* **118**:3564–3573.)

and IL-2R genes. However, IL-2 drives G_1 progression through the R-point to the S-phase. Once the cell passes the R-point, IL-2 is no longer necessary, as the cell then is irrevocably committed to initiate DNA replication, mitosis and undergo cytokinesis to yield daughter cells.

Furthermore, the Structure Activity Relationships (SAR) of the TCR versus the IL-2R now provide the discernible molecular differences in these receptor-signaling systems that lead to their relative importance to malignant transformation. Some signaling pathways, such as the Ras/Raf/MAPK pathway are shared by both the TCR and the IL-2R.[18] However, as shown in Fig. 28.2, in signaling competence, the TCR activates a distinct set of kinases and transcription factors, compared with the IL-2R, which signals cell cycle progression via the JAK/STAT system.[19] The consequence is that these two respective ligand/receptor systems activate distinct genetic programs.[20] Because TCR signaling only promotes cell cycle competence, but not cell cycle progression, it has not been found to be prominent in leukemogenesis. By comparison, the IL-2/IL-2R signaling through $JAKs_{1/3}$ and STAT5 promotes the expression of the genes that move the cell through G_1 and into the S-phase. Recent experimental evidence in murine systems have provided proof of principle that constitutively activated STAT5 can lead to CD8+ T cell lymphoma and has shown that TCR signaling is unnecessary.[21] Moreover, clinical evidence is mounting that indicates that both STAT3 and STAT5 are aberrantly activated in CD4+ Cutaneous T cell Lymphoma, as well as CD8+ anaplastic large T cell lymphoma (for review in Refs. 22, 23).

Also depicted in Fig. 28.2 are the subtle differences in signaling through heteromeric cytokine receptors versus homomeric cytokine receptors. Because the β and γ_c signaling chains of the IL-2R differ, they connect with distinct JAKs, i.e. JAK1 and JAK3 respectively. Thus, a mutation of only one of these enzymes does not confer constitutive phosphorylation and activation of STAT5. However, as shown on the right of the figure, homomeric receptors, such as employed by G-CSF, EPOP and TPO, only interact with one PTK (JAK2), so that a point mutation of this enzyme in one of the myeloid

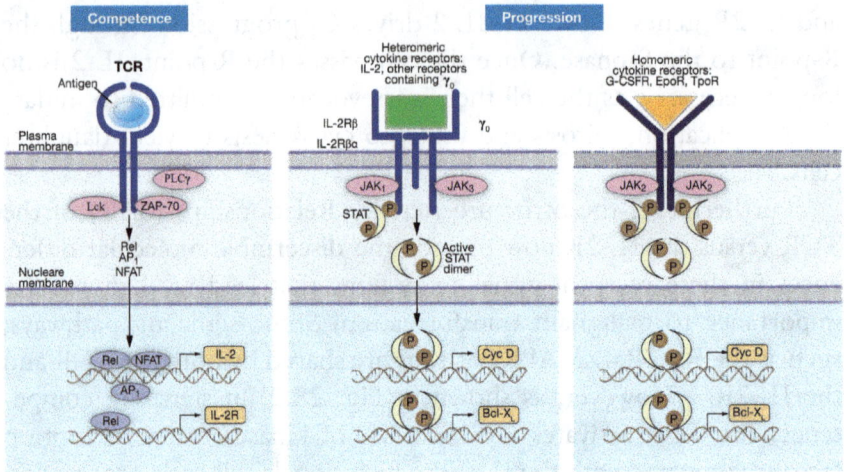

Figure 28.2: Structure-function relationships of the receptors controlling T lymphocyte and myeloid cell cycle competence and progression. The TCR activates cytoplasmic kinases (including Lck, ZAP-70, and PLCγ), which signal via intermediates to induce the activation of members of three distinct families of transcription factors, Rel, AP-1, and NFAT. These transcription factors then coordinate the expression of the genes encoding IL-2 and the IL-2Rα chain, thereby rendering the cell competent to bind IL-2 and progress through the cell cycle. Heteromeric cytokine receptors, here represented by the IL-2R, are composed of two or three distinct transmembrane chains. Binding of the cytokine to the external receptor chain domains brings the cytoplasmic domains into close enough proximity for their respective receptor-associated JAK molecules to initiate signaling. Subsequently, the activated STAT dimers translocate to the nucleus and initiate transcription of genes encoding proteins such as cyclin D_2 (Cyc D) to promote progression through the R-point, and also genes encoding cell survival proteins, such as Bcl-X_L. By comparison, homomeric receptors already have identical JAK molecules bound in close proximity, and ligand binding readily initiates signaling by promoting transphosphorylation of JAK2, the receptor chains, and eventually STAT5. (From: Smith, K.A. and Griffin, J.D. 2008. *J. Clin. Invest.* **118**:3564–3573.)

committed precursors predisposes to the constitutive activation of STAT5, and gives rise to the myeloproliferative diseases.

Accordingly, we now know *what* happens to transform a cell from normality to a malignancy. However, we do not know exactly *how* the mutational changes in critical signaling pathways affect their damage. In other words, because we have not yet discerned how cells *count* the number of cytokine hits onto the receptors, and how this information is transferred quantitatively to the signaling apparatus, the transcription factors and the genes critical to circumvent the R-point, we cannot know the molecular dynamics involved in this critical decision that the cell makes normally when it decides to divide. Furthermore, additional quantal decisions result from the initial cytokine-determined passage through the R-point. The decision to duplicate the DNA exactly, as well as the decision to segregate sister chromatids and initiate anaphase, are controlled with exquisite high fidelity. Understanding the forces underlying the molecular dynamics of these critical cellular events will serve as the quests for the future.

References

1. Smith, K.A., and Griffin, J.D. (2008) Following the cytokine signaling pathway to leukemogenesis: a chronology. *J. Clin. Invest.* **118**:3564–3573.
2. Nowell, P., and Hungerford, D. (1960) Chromosome studies on normal and leukemic human leukocytes. *J. Nat. Ca. Inst.* **25**:85–109.
3. Ponticelli, A., Whitlock, C., Rosenberg, N., and Witte, O. (1982) *In vivo* tyrosine phosphorylations of the Abelson virus transforming protein are absent in its normal cellular homologue. *Cell* **29**:953–960.
4. Klejman, A., Schreiner, S., Nieborowska-Skorska, M., Slupianek, A., Wilson, M., Smithgall, T., and Skorski, T. (2002) The Src family kinase Hck couples BCR/ABL to STAT5 activation in myeloid leukemia cells. *EMBO J.* **221**:5766–5774.
5. Danial, N., Pernis, A., and Rothman, P. (1995) Jak-STAT signaling induced by the v-abl oncogene. *Science* **260**:1875–1877.
6. Nowell, P.C. (1976) The clonal evolution of tumor cell populations. *Science* **194**:23–28.
7. Rowley, J. (1973) A new consistent chromosomal abnormality in chronic myelogenous leukemia identified by quinacrine fluorescence and Geimsa staining. *Nature* **243**:290–293.
8. Heisterkamp, N., Stephenson, J., Groffen, J., Hansen, P., de Klein, A., Bartram, C., and Grosveld, G.C. (1983) Localization of the *c-abl* oncogene adjacent to a translocation break point in chronic myelocytic leukemia. *Nature* **306**:239–242.
9. Groffen, J., Stephenson, J., Heisterkamp, N., de Klein, A., Bartram, C., and Grosveld, G.C. (1984) Philadelphia chromosomal breakpoints are clustered within a limited region, bcr, on chromosome 22. *Cell* **36**:93–99.
10. Druker, B., Talpaz, M., Resta, D., Peng, B., Buchdunger, E., Ford, J., Lydon, N., Kantarjian, H., Capdeville, R., Ohno-Jones, S., *et al.* (2001) Efficacy and

safety of a specific inhibitor of the BCR-ABL tyrosine kinase in chronic myeloid leukemia. *N. Eng. J. Med.* **344**:1031–1037.

11. Druker, B., Sawyers, C., Kantarjian, H., Resta, D., Reese, S., Ford, J., Capdeville, R., and Talpaz, M. (2001) Activity of a specific inhibitor of the BCR-ABL tyrosine kinase in the blast crisis of chronic myeloid leukemia and acute lymphoblastic leukemia with the Philadelphia chromosome. *N. Eng. J. Med.* **344**:1038–1042.

12. Yu, C.-L., Meyer, D., Campbell, G., Larner, A., Carter-Su, C., Schwartz, J., and Jove, R. (1995) Enhanced DNA-binding activity of a Stat3-related protein in cells transformed by the *src* oncoprotein. *Science* **269**:81–83.

13. Bromberg, J.F., Horvath, C.M., Besser, D., Lathem, W.W., and Darnell, J.E., Jr. (1998) Stat3 sctivation is required for cellular transformation by v-src. *Mol. Cell. Biol.* **18**:2553–2558.

14. Hoelbl, A., Kovacic, B., Kerenyi, M.A., Simma, O., Warsch, W., Cui, Y., Beug, H., Hennighausen, L., Moriggl, R., and Sexl, V. (2006) Clarifying the role of Stat5 in lymphoid development and Abelson-induced transformation. *Blood* **107**:4898–4906.

15. Yu, H., Pardoll, D., and Jove, R. (2009) STATs in cancer inflammation and immunity: a leading role for STAT3. *Nature Reviews Cancer* **9**:798–809.

16. Patinaik, M., and Tefferi, A. (2009) Molecular diagnosis of myeloproliferative neoplasms. *Expert Rev. Mol. Diagn.* **9**:481–492.

17. Quesnelle, K., Boehm, A., and Grandis, J. (2007) STAT-mediated EGFR signaling in cancer. *J. Cell. Biochem.* **102**:311–319.

18. Zmuidzinas, A., Mamon, H.J., Roberts, T.M., and Smith, K.A. (1991) Interleukin 2-triggered Raf-1 expression, phosphorylation, and associated kinase activity increase through G1 and S in CD3-stimulated primary human T cells. *Mol. Cell. Bio.* **11**:2794–2803.

19. Beadling, C., Guschin, D., Witthuhn, B., Ziemiecki, A., Ihle, J., Kerr, I., and Cantrell, D. (1994) Activation of JAK kinases and STAT proteins by interleukin-2 and interferon alpha, but not the T cell antigen receptor, in human T lymphocytes. *EMBO J.* **13**:5605–5615.

20. Beadling, C., and Smith, K. (2002) DNA array analysis of interleukin-2-regulated immediate/early genes. *Med. Immunol.* **1**:2.

21. Bessette, K., Lang, M., Fava, R., Grundy, M., Heinen, J., Horne, L., Spolski, R., Al-Shami, A., Morce, H.r., Leonard, W., *et al.* (2008) A Stat5b transgene is capable of inducing CD8+ lymphoblastic lymphoma in the absence of normal TCR/MHC signaling. *Blood* **111**:344–350.

22. Mitchel, T., and John, S. (2005) Signal transducer and activator of transcription (STAT) signalling and T cell lymphomas. *Immunology* **114**:301–312.

23. Lewis, R., and Ward, A. (2008) Stat5 as a diagnostic marker for leukemia. *Expert Rev. Mol. Diagn.* **8**:73–82.

Epilogue

This volume has focused on the molecular regulation of immune reactivity from the viewpoint of IL-2, the principle T cell growth factor responsible for the proliferative T cell clonal expansion that occurs after clonal selection. However, it is conjectured that the principles governing the molecular interactions that ultimately lead to the quantal decision on the part of the cell to progress through the cell cycle and to divide in response to IL-2 signaling are similar or identical to the intermolecular reactions that govern all of the interleukins, which function as leukocytrophic hormones. Thus, we are dealing with one of the systems of the body that is integrated with all of the other systems of the body, and consequently, internally regulated. Uncovering just how the many molecules of the immune system communicate with one another by influencing their target cells will be the task for the future. Herein lies the key to the therapeutic manipulation of the immune system, either to enhance or suppress it.

New York, New York
December, 2009

Index

www.ingramcontent.com/pod-product-compliance
Lightning Source LLC
Chambersburg PA
CBHW050555190326
41458CB00007B/2047